Smart Computing and Intelligence

This book series aims to establish itself as a medium for the publication of new research and development of innovative paradigms, models, architectures, conceptual underpinnings and practical implementations encompassed within smart computing and intelligence.

The scope of the series includes but is not limited to smart city, smart education, health informatics, smart ecology, data and computational analytics, smart society, smart learning, complex systems-chaos, computational thinking, brain computer interaction, natural/computer interaction, humanoid behaviour, and impact of educational psychology on computing.

The cornerstone of this series' editorial policy is its unwavering commitment to report the latest results from all areas of smart computing and intelligence research, development, and practice. Our mission is to serve the global smart computing and intelligence community by providing a most valuable publication service.

Ronghuai Huang · Dejian Liu · Jinbao Zhang · Yanyan Li
Editors

A Collection of AI Innovations by Chinese Teenagers

Discovering Youthful Ingenuity

Springer

Editors
Ronghuai Huang
Smart Learning Institute
Beijing Normal University
Beijing, China

Dejian Liu
Smart Learning Institute
Beijing Normal University
Beijing, China

Jinbao Zhang
School of Educational Technology
Beijing Normal University
Beijing, China

Yanyan Li
School of Educational Technology
Beijing Normal University
Beijing, China

ISSN 2522-0888 ISSN 2522-0896 (electronic)
Smart Computing and Intelligence
ISBN 978-981-97-3315-6 ISBN 978-981-97-3316-3 (eBook)
https://doi.org/10.1007/978-981-97-3316-3

This Springer imprint is published by the registered company Springer Nature Singapore Pte Ltd.
The registered company address is: 152 Beach Road, #21-01/04 Gateway East, Singapore 189721, Singapore

Foreword

In our rapidly evolving world, we stand on the threshold of a revolutionary era defined by the intellectual prowess of machines—a shining testament to human ingenuity. Artificial intelligence, a phrase once confined to the realms of science fiction, has transgressed these boundaries and become closely intertwined with the realities of our everyday lives.

In *A Collection of AI Innovations by Chinese Teenagers: Discovering Youthful Ingenuity*, we bear witness to an enlightening fusion of youthful curiosity and technological advancement. Embodied in these youthful vanguards are the seeds of a future brimming with greater potential for scientific and technological development. The exuberance they bring, combined with their mastery over AI, is reshaping our world and opening up exciting frontiers.

This book is not merely a showcase of the laudable achievements in the realm of AI by Chinese teenagers—it serves as a beacon, igniting the spark of creativity in young minds and encouraging them to venture into the exciting territory that AI presents. It is owing to the concerted efforts of numerous young trailblazers, educators, and tireless pioneers such as the Yuanzhuo Initiatives, that this beacon could be lit.

The innovative pursuits encapsulated in this collection, originating from various corners of China, span a broad spectrum of applications—from endeavoring to secure the foothold of the vastly diverse avian population to ensuring food safety and elevating the quality of our daily lives. Their work is a testament to the versatility of AI, a tool that can be wielded aptly by well-trained young hands to revolutionize an array of fields.

It is indeed laudable to see how their innovative projects, such as the Swift Population Monitoring system, demonstrate the power of AI models to make a significant societal contribution. Equally remarkable is the contribution from Macao, which has revealed the vast potential of AI in enhancing the reliability of machines. These inspirational examples bear evidence to the impetus AI has provided to teenagers in reshaping the world.

This collection of cases holds a dual edge. On the one hand, it aims to inspire more youth to tread on the path of exploratory learning and encourage AI applications in the quest to overcome real-world challenges. On the other hand, it offers resources for education hubs, thereby broadening the horizon for AI research considerably.

In closing, the melody of AI continues to resonate and grow louder. The harmony articulated in this book allows us to appreciate the passionate efforts of Chinese teenagers. Let this symphony of youthful innovation stir in us a renewed zeal for pushing boundaries and spearheading discovery. Their exploration spirit, innovation awareness, and technical prowess acted as guiding stars in their grand journey of AI. And so, with great anticipation, I invite you to delve into these pages and lose yourself in their world—for their journey today shapes the road we tread tomorrow.

Lei Fan
Professor
Capital Normal University
Beijing, China

Preface

In the twenty-first century, an era full of challenges and opportunities, artificial intelligence has become an important driving force for developing science and technology. With the progress of technology and the widespread practical applications, artificial intelligence has gradually penetrated all aspects of our lives, thus shaping a whole new world. In this transformation, the youth generation is responsible for contributing to future innovation and development.

UNESCO issued the *Beijing Consensus on Artificial Intelligence and Education* in 2019. In order to respond to the call and to show the outstanding achievements as well as the exploration spirit in the youth artificial intelligence field in China, therefore promoting the development of youth artificial intelligence education, Yuanzhuo Initiatives launched the first *Solicitation of Outstanding Artificial Intelligence Cases* in 2022. The solicitation collected projects focusing on real problems by using original and innovative algorithms of artificial intelligence by students in secondary schools in China. A total of hundreds of cases were received during the event, 10 of them were shortlisted after 3 rounds of selection and defense at various levels. The projects come from Beijing, Shanghai, Macau, Chengdu, Fuzhou, and other places. The fields and topics involved are different, including industry, medicine, food security, express delivery, helping the elderly, smart home, and other fields.

Behind these cases are a group of enthusiastic teenagers who pursue excellence. They have not only learned and applied artificial intelligence technology but also endeavor to solve practical problems in life, bringing real changes and progress to society. In these innovative cases, we can see the exploration spirit, innovation awareness, and technical strength of Chinese teenagers in the field of artificial intelligence, such as the project from Beijing Middle School: *A Monitoring System for Migratory Bird Beijing Swift with Object Detection*, which assisted Beijing Wildlife Rescue and Rehabilitation Center and solved the problem of low efficiency of volunteers' manual marking of swifts by using deep learning image processing technology, and achieved very good results in practice. Another example is *A Rotor Fault Detection System Based on Nonlinear and Dynamic Response* from Lou Hau High School, Macao, which focused on motor rotor fault detection. A detection rate of nearly 100%

has been achieved through continuous improvement of algorithms and engineering practice.

We hope that the compilation and distribution of this casebook can provide more learning references and enlightenment for teenagers, stimulate their concern and passion for the field of artificial intelligence, increase interest in learning, and improve learning ability, therefore promoting the development of artificial intelligence education for teenagers. Meanwhile, this book also provides references and resources for artificial intelligence education and research institutions, contributing to China's scientific and technological innovation and development. Therefore, we expect these excellent cases to receive more attention and recognition in the field of artificial intelligence and jointly promote the development of artificial intelligence globally.

Ronghuai Huang
Chair-Holder, UNESCO Chair on AI in Education

Professor, Beijing Normal University
Beijing, China

Introduction—Youth Artificial Intelligence Innovation Initiatives: The Yuanzhuo Initiatives for Chinese Teenagers

Background

Artificial intelligence (AI) is driving the rapid development of ICT in education, and AI in education has become a dynamic and diverse field with significant advancements. AI has been widely applied to various aspects of education, including personalized learning, adaptive assessment, intelligent tutoring, curriculum design, teacher training, and educational administration. AI has created value for students, teachers, schools, and regional education systems by improving teaching quality, efficiency, and equity.

According to UNESCO, AI presents multiple risks and challenges, such as ethical, legal, and social implications, necessitating policy debates and regulatory frameworks. Accordingly, UNESCO has been spearheading several initiatives to support Member States in harnessing the potential of AI for achieving the Education 2030 Agenda while ensuring that the core principles of human rights and human dignity guide its application in educational contexts. Some of these initiatives include:

- *The Beijing Consensus on Artificial Intelligence and Education* established a shared vision and action plan to promote AI in education (UNESCO 2019).
- The publication *Artificial Intelligence and Education: Guidance for Policy-makers* provided practical recommendations and examples for developing and implementing AI policies in education (UNESCO 2021).
- In March 2023, the International Forum on Artificial Intelligence and Digital Education, together with the Establishment Ceremony of the UNESCO Chair on Artificial Intelligence in Education, takes place at Beijing Normal University; the objective is to facilitate a platform that enables the exchange of knowledge and experiences among professionals, educators, and policymakers who possess expertise in the domain of AI in education.

Throughout this process, governments and relevant organizations worldwide have implemented various policies and standards to encourage the integration and growth of artificial intelligence within education. For example, in April 2022, Information

Technology Curriculum Standards for Chinese Compulsory Education were introduced, incorporating AI into the learning content of grades 7–9 in China's compulsory education system. Furthermore, in 2019, the National Science Foundation (NSF) of the USA initiated the National Artificial Intelligence Research Institutes Program to establish a network of interdisciplinary and collaborative institutes to advance AI research and education. Additionally, the European Union (EU) launched the Digital Education Action Plan 2021–2027, aimed at fostering digital skills and competencies for digital transformation while enhancing the digital capacity of education and training institutions.

The teenage years present the optimal opportunity to nurture interests and uncover potential, as this is when their thinking, cognition, and innovation abilities rapidly develop. Incorporating AI education during adolescence is vital for cultivating talents with an innovative spirit and the ability to adapt to future societal needs. Introducing AI education during this period aids in developing innovative, logical analysis skills and problem-solving abilities. Engaging with and learning about AI from a young age can inspire students' interest in technology and innovation, helping them to better adapt to future societal demands and enhance their competitiveness. Moreover, AI education requires students to acquire interdisciplinary knowledge and skills, offering them opportunities to improve their comprehensive capabilities.

Establishment of Youth Artificial Intelligence Innovation Initiatives

Beijing Normal University (BNU), a leading university for teacher education in China, has pioneered advancing AI in education and research. With its rich academic excellence and innovation history, BNU acknowledges the significance of fostering future leaders in the AI domain. The university has established numerous research centers, laboratories, and interdisciplinary programs to offer comprehensive training and resources for students, teachers, and researchers interested in AI. In November 2022, UNESCO approved the establishment of a UNESCO Chair on Artificial Intelligence in Education at BNU, signifying enhanced connections between research, policy, and practice at country, regional and global levels.

BNU's commitment to fostering talent and driving innovation has resulted in the creation of the Youth Artificial Intelligence Innovation Initiatives, as known as Yuanzhuo Initiatives, a pioneering initiative in AI education for youth. In December 2019, BNU launched Yuanzhuo Initiatives in response to the call for implementing the *Beijing Consensus on Artificial Intelligence and Education.* This program seeks to advance AI education among youth by fostering collaboration between academia, research institutions, and industry. Our goal is to cultivate top-notch innovative youth talents with the ability to apply original and groundbreaking algorithms to address real-world challenges.

Origin of the Title YUANZHUO

Historical records indicate that LIU Hong, also known by the courtesy name Yuanzhuo, was the inventor of the Chinese abacus (Suan-Pan in Chinese). Living during the Liu Song Dynasty in the Southern and Northern Dynasties period of ancient China, around the 5th century AD, Liu was a distinguished mathematician, astronomer, and engineer. In his work "Qian Xiang Li" (Uranic Manifestation Calendar System), he presented a comprehensive account of the primary irregularities in lunar motion. This account was based on a detailed analysis of the vast data collected by Han dynasty sky-watchers over the preceding centuries, enabling solar eclipses' prediction. Liu's contributions led to the abacus becoming the foremost calculating instrument in ancient China, greatly impacting the advancement of mathematics, commerce, and science.

Hence, the name YUANZHUO was chosen for the Youth Artificial Intelligence Innovation Initiatives (YAIII), drawing inspiration from LIU Hong's courtesy name to represent originality and excellence. This embodies the nurturing of innovative spirit and exceptional talents in youth, paving the path for the forthcoming era of artificial intelligence.

Major Projects of Yuanzhuo Initiatives

(1) **Yuanzhuo Resource Platform of Algorithm, Datasets, Courses, and Textbooks**
Launched in 2019, the Yuanzhuo Resource Platform (https://yuanzhuo.bnu.edu.cn) now offers nearly ten thousand types of artificial intelligence resources. This includes renowned teachers' lectures, expert reports, thematic seminars, artificial intelligence datasets, open computing resources based on JupyterHub, Yuanzhuo Initiatives self-developed open-source courses, and various academic resources provided by the National Engineering Research Center for Cyberlearning and Intelligent Technology (https://cit.bnu.edu.cn) and the Smart Learning Institute of Beijing Normal University (http://sli.bnu.edu.cn). The platform has received millions of visits from over 70 countries and regions worldwide.

(2) **Yuanzhuo Community for ICT Teachers and Students**
The Yuanzhuo community embraces open source and mutual support. Experts, teachers, and students participate in community activities online and offline, including the official website, WeChat groups, Tencent meetings, and public account live broadcasts. Among these, Yuanzhuo Academy serves as an online communication platform for the community and has successfully hosted over forty online exchange activities since 2022. In addition, over one hundred esteemed guests have shared valuable insights on various topics, including

discussions on curriculum standards, presentations on cutting-edge innovations such as ChatGPT, and the metaverse, teaching methodologies, and regular reading undertakings, eliciting enthusiastic responses from community members.

(3) **Global Youth Cooperation Through AI-Themed Summer or Winter Schools**
To promote cross-border cooperation and communication among young talents from diverse countries, we organize global youth AI-themed summer camps that provide invaluable guidance from experts from China and abroad. These camps serve as a platform for fostering international collaboration and nurturing the next generation of global leaders in the field of artificial intelligence. In the summer of 2022, together with the Computational Thinking Research Center of Beijing Normal University, we held an academic forum on the theme of computational thinking education across the Taiwan Strait and a summer camp for students with more than 400 participants.

(4) **Boosting Case Studies by Call for Youth Contributors**
The Yuanzhuo Initiatives actively identify and commend outstanding achievements of young talents who demonstrate exceptional prowess in utilizing original and innovative AI algorithms to solve real-world problems. As a result, selected projects will be given the opportunity to be showcased internationally, elevating the status of China as a prominent global innovation hub for artificial intelligence.

Youths participating in Yuanzhuo Initiatives who demonstrate promising achievements are encouraged to explore research-worthy problems in artificial intelligence with comprehensive guidance and support. We are committed to providing extensive assistance through a series of community activities. Online community platforms that offer access to algorithms, computing power, datasets, knowledge, and invaluable insights to foster excellence among young talents continuously. Information technology teachers play a crucial role as instructors in enhancing students' technical knowledge and helping throughout the project journey. Experts from research institutions guide students to appreciate the fascinating aspects of AI technology and algorithms from a scientific perspective. Engineers from technology companies contribute by sharing cutting-edge algorithm cases that address practical challenges faced by enterprises and help students overcome technical hurdles.

The Yuanzhuo Initiatives launched the first Solicitation of Outstanding Artificial Intelligence Cases of Chinese Youth in 2022. The project collected projects about real problems by using original and innovative algorithms of artificial intelligence by students in secondary schools in China. Hundreds of cases were received during the event, with ten projects shortlisted after rigorous selection and defense at various levels. The projects originated from locations such as Beijing, Shanghai, Macau, Chengdu, and Fuzhou, covering diverse fields and topics, including industry, medicine, food security, express delivery, elderly assistance, waste sorting, smart homes, and smart classrooms.

Under the guidance and support of UNESCO and relevant organizations, we will initiate a campaign to collect outstanding youth achievements in artificial intelligence across more than 10 countries worldwide. A compilation of these cases is set to be released in the summer of 2024, showcasing the accomplishments of youth worldwide in the field of artificial intelligence.

The Yuanzhuo Initiatives are committed to promoting AI education for youth through cooperative approaches, nurturing innovation, and supporting exceptional accomplishments. We are confident that by equipping young talents with essential knowledge, skills, and resources, we can contribute to the progress of artificial intelligence for all.

Features of This Book

(1) **Youthful Perspective**

As the title suggests, *Youthful Ingenuity* offers a unique perspective on AI from the younger generation. The book reflects the fresh and innovative mindset of teenagers growing up in the era of rapid technological advancements. Their unique approach, creativity, and out-of-the-box thinking bring a fresh perspective to the field of AI. This youthful perspective adds a dynamic and refreshing element to the book, making it an engaging read for experts and novices in AI.

(2) **Innovations by Chinese Teenagers**

The most distinguishing feature of *Youthful Ingenuity* is that it presents innovations by Chinese teenagers, the driving force behind the book. These young minds have produced cutting-edge ideas, projects, and applications that demonstrate their creativity, curiosity, and determination to be effective in the field of AI. The book provides a platform for these teenagers to showcase their groundbreaking work, which is impressive and inspiring.

(3) **AI-Focused Content**

Youthful Ingenuity stands out for its exclusive focus on AI-related content. The book covers various topics, including machine learning, computer vision, natural language processing, robotics, and more. It delves into the technical details of the innovations, providing readers with a comprehensive understanding of the concepts and technologies involved. The book also discusses the challenges young innovators face and their approaches to overcoming them, making it a valuable resource for those interested in the practical applications of AI.

(4) **Diverse Topics Covered**

Another distinctive feature of *Youthful Ingenuity* is the wide array of topics. The book showcases various AI projects and innovations, ranging from healthcare and education to agriculture and transportation. This diversity of topics demonstrates the versatility and applicability of AI in various domains and provides readers with a holistic view of the potential of AI in addressing real-world challenges.

(5) **Practical Applications of AI**
Youthful Ingenuity emphasizes the practical applications of AI in solving real-world problems. The book focuses on how AI can address pressing issues such as disease diagnosis and environmental sustainability. The book provides practical insights, case studies, and examples of how these young innovators have leveraged AI to create meaningful impact and contribute to the betterment of society. This emphasis on real-world applications makes *Youthful Ingenuity* a relevant and valuable resource for those interested in the practical implications of AI.

Contents

Chapter 1
A Monitoring System for Migratory Bird Beijing Swift with Object Detection

He Li, Zixuan Wang, and Manming Zhao

Abstract This project outlines the development of an intelligent assistant application (APP) for the Beijing Swift Survey Project in 2022, which uses advanced data collection and processing techniques. The system employs an object detection algorithm to monitor and count swifts accurately, improving the accuracy of swift counting processes while streamlining data collection. Furthermore, the system utilizes the Internet for increased publicity and aims to enhance the user experience through continuous optimization and innovative features. The Beijing Swift Survey Project is to assess the conservation status of swifts in Beijing and is an essential component of the city's biodiversity conservation efforts. The use of artificial intelligence can significantly improve the efficiency and accuracy of data collection, which can have far-reaching implications for ecological research and conservation efforts. The development of this APP represents a promising solution to challenges faced in swift monitoring, and it can potentially become a model for monitoring other bird species in the future.

Keywords Beijing Swift · Object detection · Web application (APP)

Beijing Academy was established in September 2013 and approved by Beijing Municipal Government. Adhering to the orientation of internationalization, modernization, and high quality, the school is committed to building a modern school with Beijing style, Chinese temperament and world mind, and gradually developing into a world-class Chinese famous school.

H. Li · Z. Wang · M. Zhao (✉)
Beijing Academy, Beijing 10028, China
e-mail: swellfishming@163.com

H. Li
e-mail: li@imlihe.com

Z. Wang
e-mail: henrywang3510@gmail.com

R. Huang et al. (eds.), *A Collection of AI Innovations by Chinese Teenagers*, Smart Computing and Intelligence, https://doi.org/10.1007/978-981-97-3316-3_1

Highlights

This project outlines the development of an intelligent assistant application that uses advanced data collection and processing techniques to accurately monitor and count swifts. With the use of artificial intelligence, this project represents a promising solution to challenges faced in swift monitoring and can potentially become a model for monitoring other bird species in the future.

Challenges The expenses associated with tagging swallows significantly escalate the cost of acquiring extensive, effective training data. Current deep learning methods for target detection demonstrate limited accuracy when identifying small targets. Furthermore, the precision of these existing techniques continues to be inadequate for small target detection.

Solution The project's dataset was primarily developed through a combination of manual tagging and pseudo-labeling generated by the model.

Outcome The deliverables comprised a deep learning model and an integrated web system (inclusive of a website and a mini program).

What is the Beijing Swift Survey Project about and why is it important for biodiversity conservation efforts?

The Beijing Swift Survey Project is a project aimed at assessing the conservation status of swifts in Beijing. It is an essential component of the city's biodiversity conservation efforts. The use of artificial intelligence can significantly improve the efficiency and accuracy of data collection, which can have far-reaching implications for ecological research and conservation efforts.

How does the object detection algorithm improve the accuracy of swift counting processes?

The object detection algorithm employed by the system can better highlight the features of the tiny target, thereby enhancing the detection accuracy without significantly decreasing the detection speed. The improved CenterNet model has been tested and shown to have a significant performance and speed improvement compared with the pre-improvement, meeting the requirements of the target detection system in the project objectives.

How does the use of artificial intelligence in bird monitoring have far-reaching implications for ecological research and conservation efforts?

The use of artificial intelligence in bird monitoring can significantly improve the efficiency and accuracy of data collection, which can have far-reaching implications for ecological research and conservation efforts. By accurately monitoring and counting bird populations, researchers can better understand the ecological health of an area and make informed decisions about conservation efforts. The development of the intelligent assistant application for the Beijing Swift Survey Project represents a promising solution to challenges faced in swift monitoring, and it can potentially become a model for monitoring other bird species in the future.

1.1 Overview of Beijing Swift Survey Project

1.1.1 Background

Beijing Swift, scientifically known as the hawk swallow, stands as an avian wonder within the realm of Caprimulgiformes, a fascinating order of birds that includes the highly agile Apodidae. This charismatic bird, aptly named after the bustling metropolis of Beijing, exhibits a plethora of distinctive biological traits that set it apart from the avian crowd. However, the relentless march of urbanization and the demolition of ancient architectural treasures have cast a shadow on the once-thriving population of Beijing Swifts. These birds, renowned for their historic preference to nest beneath the eaves of time-worn structures, have faced the brunt of urban development.

In a commendable effort to safeguard the dwindling Beijing Swift population, the Beijing Wildlife Rescue Center embarked on the ambitious Beijing Swift Scientific Survey Project back in 2017. With the aim of enlisting the aid of dedicated volunteers from the public, this noble initiative sought to rally forces for the conservation of these splendid creatures. It was in 2018 when the call for volunteers resonated through the city, and individuals from various walks of life enthusiastically joined the cause.

The survey itself presented a unique set of challenges. These swift, small, and nimble beings are known for their lightning-fast flight patterns, often preferring to soar through the skies in clustered formations. This distinctive behavior posed a considerable challenge to the volunteers tasked with documenting their flights. Yet, despite the hurdles, the project persisted, spurred on by an ever-growing team of passionate individuals.

As the project continued to gain momentum, an astonishing surge in social volunteers lent their expertise to the study. The number of these dedicated souls swelled to nearly 500, forming a formidable force of conservation enthusiasts. Consequently, the early stages of the project, with just a handful of survey points, blossomed into a sprawling network of nearly 100 active survey locations. The sheer volume of survey data generated by this army of volunteers presented an unexpected and substantial challenge.

To tackle the dual challenges of documenting the elusive Beijing Swift and managing the burgeoning data streams, the project's coordinators embarked on a journey of development and enhancement. The objectives of this evolution in the project were as follows:

1. To refine and enhance the survey methods, making them more adaptable to the unique flight patterns and behaviors of the Beijing Swift.
2. To establish a robust framework for data collection and management, ensuring that the wealth of information collected by the growing number of volunteers is effectively processed and analyzed.
3. To promote public awareness about the conservation needs of the Beijing Swift and engage even more citizens in the cause.

4. To forge partnerships with relevant stakeholders, including local authorities, to ensure the protection of critical habitats for the Beijing Swift.
5. To foster a sense of community and shared responsibility among volunteers, further strengthening their commitment to the preservation of this remarkable species.

In summary, the Beijing Swift Scientific Survey Project is not merely a data-gathering endeavor but a passionate and evolving movement driven by the dedication of volunteers and the commitment of those who understand the vital role these birds play in the urban ecosystem. Through perseverance, collaboration, and innovation, this project aspires to secure a brighter future for the Beijing Swift amidst the ever-changing landscape of Beijing's urban sprawl. The project aims to use artificial intelligence and network technology to solve the difficulties encountered in the Beijing Swift Survey project and to support further development. The specific objectives are:

(1) Using deep learning image processing technology to complete photograph annotation quickly.
(2) Build a system to aggregate survey data and add more functionality.
(3) Build a portal website for the survey project to help promote it.

The project will fully use of the network and artificial intelligence technology to facilitate the Beijing Swift Survey Project volunteers. The project team should also fully understand the needs and opinions of volunteers and survey project leaders, continuously optimize the iterative system and introduce more functions.

1.1.2 Why to Monitor Beijing Swift

This innovative project represents a commendable fusion of cutting-edge technologies with the noble cause of preserving the Beijing Swift population. By harnessing the power of artificial intelligence and Internet technology, it has not only streamlined the Beijing Swift Survey Project but also opened up exciting possibilities for conservation efforts on a broader scale.

In the current landscape, the system has garnered positive feedback, demonstrating its effectiveness in enhancing the efficiency of volunteer surveys. This leap in efficiency is undoubtedly a significant stride forward in ensuring the preservation of these remarkable birds. As a testament to its success, the system has standardized the data collection approach for the Beijing Swift Survey Project in 2022, setting a new precedent for the project's future endeavors.

Moreover, the project's visionary outlook extends beyond the realm of data management. It seeks to leverage the vast potential of the Internet for outreach and awareness campaigns, amplifying the message of conservation for the Beijing Swift and potentially inspiring others to join the cause. This forward-thinking approach not only consolidates the project's impact within the local community but also holds promise for similar initiatives worldwide.

The project team's commitment to continuous improvement is evident in their dedication to optimizing user experiences and developing novel functions. This adaptability and pursuit of excellence ensure that the project remains on the cutting edge of technological advancements, ultimately benefiting the avian species it aims to protect.

In the course of achieving its objectives, the project has surmounted significant challenges, some of which hold universal applicability. A noteworthy achievement is the project's solution to the daunting problem of accurately and swiftly detecting small targets. This breakthrough not only enhances the project's efficiency but also offers valuable insights and solutions for analogous issues in various domains.

Furthermore, the project has amassed a comprehensive dataset for Swift target detection, which serves as a valuable resource for evaluating the performance of micro-target detection models. This data repository not only advances the project's own objectives but also provides a foundation for future research and development in the field of object detection and recognition.

In summary, the true value of this project shines through its unwavering support for the Beijing Swift Survey Project and its pioneering solutions to pertinent technical challenges. As it continues to forge ahead, this initiative stands as a beacon of hope, showcasing how technology, when harnessed for a noble cause, can pave the way for a brighter future not just for the Beijing Swift but for countless other species in need of conservation and protection.

1.2 Technological Issues and Solutions for Beijing Swift Survey

1.2.1 Using Object Detection for the Survey

The project's main technical problem is accurately detecting swifts in the pictures. The difficulty of this problem is that swifts are often densely distributed in the picture, and each swift occupies a very tiny area. For most existing target detection methods, detecting a tiny target in a high-resolution image can lead to significant result errors or a long calculation time.

Table 1.1 shows the results of training the wheel on the annotated dataset using the currently more mature and mainstream YOLOv5 model. The inference time-consuming is the time needed for the model to completely detect a picture with resolution (including the steps of preprocessing, post-processing, formatted output, etc.). The degree of accuracy represents the percentage error between the number of detection frames with a higher threshold and the number of actual detection frames in the output result of the model. The calculation method is as follows:

$$\%\text{err} = \frac{|N_{\text{true}} - N_{\text{result}}|}{N_{\text{true}}} \times 100\%.$$

Table 1.1 Result analysis of existing target detection methods

Picture number	Reasoning time (s)	Accuracy rate (%)
37	10.7	78.5
43	11.3	68.3
86	9.8	89.1

Figure 1.1 (partial of pictures No. 37, 43, and 86 from left to right) shows the types of false detections included in the detection results output by the YOLOv5x model for three test pictures, which are error detection, missing detection, and re-detection from left to right. Among them, the number of missed tests is the most, leading to a large deviation between the model detection and real results. Rechecks and missed checks often occur in the cluster area. At the same time, error detections are reflected in the model's inability to distinguish swifts from other objects in the picture (such as leaves and eaves)

The Beijing Swift Survey Project, while making remarkable strides in its mission, faces a pivotal challenge that lies at the heart of its success: the development of a target detection system capable of swiftly and accurately identifying Beijing Swifts in images. This challenge is of paramount importance, as it directly impacts the efficiency of the project and its ability to reduce the manual workload for volunteers. Striking the right balance between technological accuracy and volunteer involvement is key.

Recognizing this, the project's second objective is equally pivotal. It seeks to harness the power of network technology to standardize the data uploading and reporting process. However, the feasibility of building such a network system must be carefully considered, with a strong emphasis on user-friendliness. It is important to bear in mind that many volunteers in the Beijing Swift Survey Project are primarily bird enthusiasts and photography aficionados, not necessarily seasoned Internet users. Therefore, the system's user interface must be designed for simplicity and accessibility to ensure a smooth experience for all volunteers.

The second challenge that the project faces revolves around the creation of this network system, which needs to possess key attributes: a quick and straightforward setup, user-friendly navigation, and a comprehensive suite of functions. In essence,

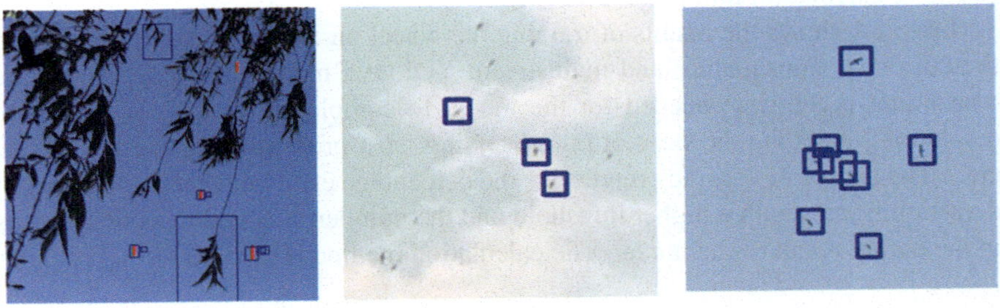

Fig. 1.1 False detection types of existing target detection methods

the project must answer the question of how to construct a network infrastructure that offers a low barrier to entry, making it easy for volunteers to report their findings promptly and effectively.

To address these intertwined challenges, the project can consider several strategies:

1. Advanced machine learning algorithms: Invest in the development and refinement of machine learning algorithms tailored specifically for swift detection. This involves continuous training and fine-tuning of the model to reduce false positives and enhance accuracy.
2. User-centric design: Collaborate with user experience (UX) designers to create an intuitive and visually appealing user interface for the network system. Prioritize simplicity in data uploading and reporting processes, ensuring that even those with limited Internet experience can use it with ease.
3. Comprehensive training: Provide comprehensive training materials and resources for volunteers to acquaint them with the network system's features and functionalities. Conduct workshops or webinars to ensure volunteers are comfortable with the technology.
4. Feedback mechanism: Implement a feedback mechanism within the network system so that volunteers can report any issues or difficulties they encounter. Use this feedback to continuously improve and optimize the system.
5. Regular updates: Commit to regular updates and maintenance of the network system to fix bugs, add new features, and enhance overall performance.
6. Community support: Foster a sense of community among volunteers, where experienced users can offer guidance and support to newcomers, making the learning curve less steep.

In conclusion, the Beijing Swift Survey Project is not just about avian conservation; it's also a testament to the harmonious integration of technology and community involvement. By surmounting the challenges of accurate target detection and user-friendly network systems, the project can inspire and empower volunteers from all walks of life to actively participate in the preservation of the Beijing Swift population, setting an example for similar initiatives worldwide.

1.2.2 Improving the Detection Accuracy

1.2.2.1 Optimized Object Detection

For the first problem, detection accuracy is the primary optimization goal. At present, there are several ways to optimize small target detection:

(1) At the model level, add the capability of multi-scale input or multi-scale feature fusion to the model.

(2) During the training of the collected data, data enhancement is applied to the training data to provide multi-scale feature information.
(3) Use the anchor-free solution.

The former two solutions focus on enhancing the models' multi-scale information processing capability. This capability also largely determines a model's ability to detect small targets.

Figure 1.2 is a typical image pyramid sampling method. In the training process, using this method to enhance the data can effectively increase the scale richness of the training set data. At the same time, models such as NAS-FPN also apply the image pyramid mechanism to the input part of the model. However, adopting multi-scale input can also cause limitations, including significantly slowing the inference speed.

Figure 1.3 shows the basic architecture of the multi-scale feature prediction fusion model. A representative of this model type, single-stage headless face detector (SSH), uses multiple convolution and pool layers to extract features of different scales, detect them separately, and finally combine their detection results. Thanks to this feature, the model outperformed then-leading ResNet-101 and improved reasoning speed by a factor of five.

Fig. 1.2 Image pyramid

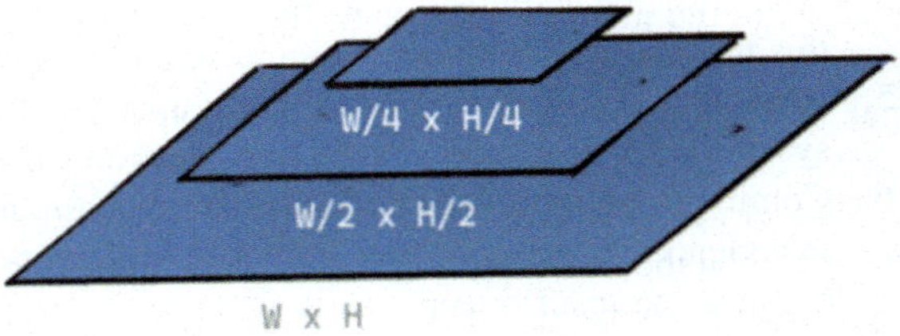

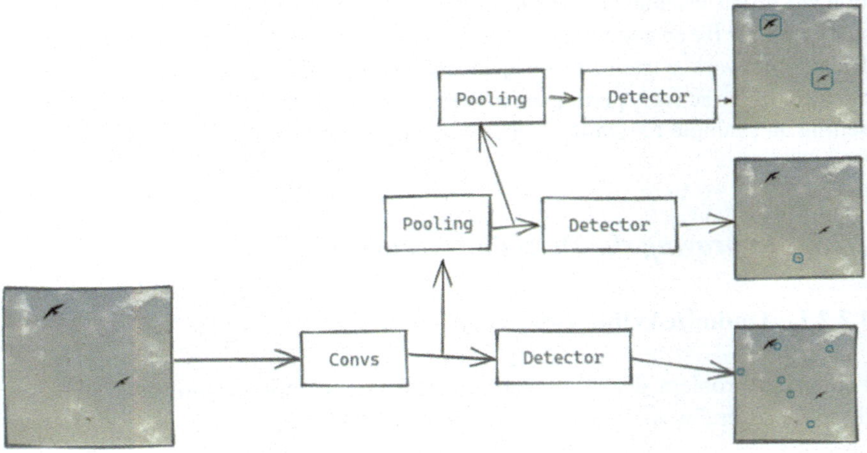

Fig. 1.3 Multi-scale feature prediction fusion target detection

Fig. 1.4 Schematic diagram of heat map in CenterNet

With the development of target detection technology, anchor technology is adopted in the mainstream and high-performance models. Although this method's performance and detection speed are significantly improved compared with the original sliding window detection method, when detecting small objects based on an anchor detection model, some of them cannot be covered entirely, which also leads to the fact that such anchor-based models tend to behave less well on small object datasets than the traditional way of using image pyramids and sliding windows.

In contrast to the anchor-based method, some anchor-free models, which are not based on the anchor technique, such as CornerNet and CenterNet, have emerged recently, of which the performance of CenterNet is excellent. The method dramatically reduces the complexity of the model and only includes one backbone network and three convolution neural networks for processing and outputting the predicted values. The backbone network can output a heat map in which the value of the target center point is larger. Therefore, each target has only one anchor point in the center, as shown in Figure 1.4.

To sum up, the problem of small target detection can be solved by three methods. At the same time, a more innovative way could be used to solve the problem by combining the advantages of multiple approaches. In this project, the whole target detection model adopts the architecture of CenterNet, in which a modified ResNet-18 replaces the original backbone network. The modification of this part mainly refers to the Hourglass network and gives the model to incorporate multi-scale features by adding an upsampling layer.

Considering that the direct use of image pyramid input will seriously affect the reasoning speed of the model and that the immediate use of image pyramid will result in the features of some smaller targets not being highlighted well, the project does not directly adopt image pyramid to realize multi-scale feature input. Like the more traditional target detection methods, the model in this project uses a sliding window

Table 1.2 Result analysis of improved target detection methods

Picture number	Reasoning time (s)	Accuracy rate (%)
37	5.2	98.6
43	4.6	99.1
86	4.3	100.0

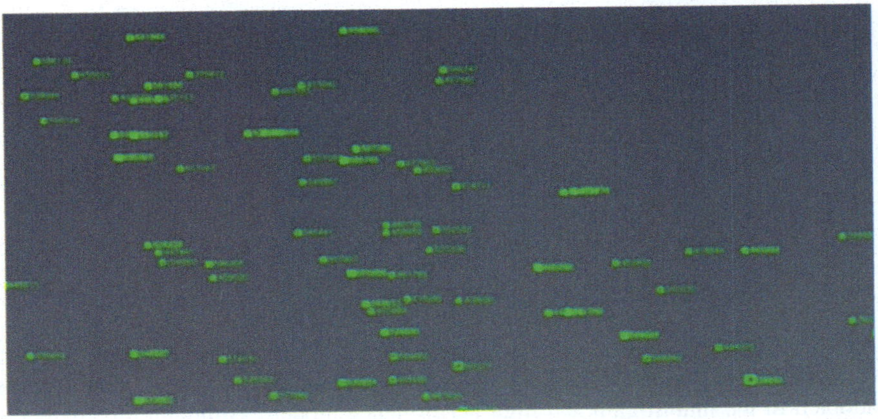

Fig. 1.5 Partial detection results of pictures

to process the input image data. This method can better highlight the features of the tiny target, thereby enhancing the detection accuracy without significantly decreasing the detection speed.

Table 1.2 shows the test results after 300 rounds of training on the dataset using the improved CenterNet model. The data measurement method is the same as in Table 1.1. By comparing the results of three test pictures, it can be proved that the improved method has a significant performance and speed improvement compared with the pre-improvement. It meets the requirements of the target detection system in the project objectives.

The labeled picture shown in Fig. 1.5 is a part of the detection result of image NO. 86. It can be found that the improved target detection scheme also has a higher detection accuracy rate for the cluster of Beijing Swift.

1.2.2.2 Simple the User Interaction and Stable Network System

The second problem mainly involves the user interaction part of the system, and the solution is to simplify the interaction design. Common ways are to remove unwanted distractions, provide instructional elements, and so on. The user interface design of this project's network system follows a modern and concise style.

On the other hand, it is vital to ensure the stability of the network system. According to the activity routines of the Beijing Swift, volunteers often must choose to observe and report at sunset every afternoon. Moreover, because most volunteers do not have time to fill in during the working day, the number of uploads on weekends will be significantly higher than on a working day. Therefore, the system must be stable during the peak reporting period, which requires the stability and speed of the target detection system. To guarantee the stability of the detection system, it is necessary to design the queue mechanism to prevent instability and even collapse caused by excessive detection tasks. Furthermore, the stability of the filling and uploading part of the system can be guaranteed by improving the back-end performance. Specific solutions include using languages with more optimized space and introducing asynchronous and multithreading in the IO operation part of the network.

1.3 Phases of Project Implementation

Sections 3.1 and 3.2 of the project implementation involve the development and iterative design of the monitoring system. In Sect. 3.1, the team creates three components to develop the monitoring system. In Sect. 3.2, iterative designs are implemented to continuously improve the monitoring system based on feedback and evolving project requirements. This combination ensures an adaptable and effective monitoring process throughout the project.

1.3.1 Development of the Monitoring System

The development of the monitoring system consists of three components. The front-end component focuses on user interface design, while the target detection component identifies and tracks desired targets. The back-end component handles data storage and processing. Together, these components create an integrated monitoring system that enables efficient monitoring, analysis, and decision-making of Beijing Swift Survey.

Front-End Component

The system's front end was built using the Vue2 framework and the ElementUI component library in the early experimental phase. Because this version is mainly for the test, the project's overall development is primarily based on the development speed, ignoring the performance, aesthetics, and other factors. Therefore, a mature front-end development framework is selected for the front-end development of this version of the system. The burden of the basic element's development is relieved with the help of the existing third-party component library. Since this system version only has a target detection function and does not involve complex interactions such as user system and data filling, the front-end part can be presented on a simple single page. This page contains only one component for uploading the pictures taken and one component for showing the detection results after the test is completed.

Because it does not involve complex operation logic, the front-end development of the system does not need to consider the running performance and maintainability. This version's front-end development started in April 2021 and was completed and launched in June 2021.

In the next release phase of the project, there will be more supplements to the functions and requirements of the system. This system version needs to involve the features of fill-in and upload and requires the system to include a user and group system so that the fill-in records of users can be traced. At the same time, this system version also involves a simple authority and management system. Project leaders and team leaders with greater authority need a way to manage the filling data and system data.

Implementation section: The front end of this version uses the updated Vue3 framework and NaiveUI component library. Earlier versions of the Vue3 front-end framework introduced the concept of a combined interface, allowing code with similar or identical functionality to be reused between components, which makes the framework ideal for building larger front-end projects. NaiveUI combines ease of use and flexibility. The elements of the component library can easily be customized to fit a particular scenario without modifying the code of the entire library or re-implementing a similar component.

Due to the expansion of the functions, the front-end portion of the release is no longer suitable for presentation as a single page. Before development, the project team first communicated with the responsible team of the Beijing Swift Survey Project to define all the functional requirements and assigned a page address for each function. In the development process, each page corresponds to a file-based component, and each page component can be introduced with shared components using other locations.

The increased complexity of the function also requires that the performance of the front-end system must be optimized. For some time-consuming operations that significantly impact the user experience, such as network requests and high-density operations, the project team adopts optimization methods such as caching to speed up page performance and simplify part of overly complex script logic. In addition, when front-end projects are packaged, all files are compressed to speed up page access.

This version of the front-end system uses a responsive layout and page design that allows each page to fit both the desktop and the client without having to develop a separate page for each client.

The next version of the front end will need to pay more attention to the ease of user interaction and the holistic aesthetics of the front-end project. As the number of functions increases, the page structure and logic of the project need to be redesigned. The new version of the front-end project will be split into several sub-projects, each deployed under a separate sub-domain.

At the same time, the front-end portion of the new release will be based on the higher-performance solids framework and will no longer use third-party component libraries. As a result, the front-end portion of this release is expected to exceed the previous two versions in terms of completion and ease of use.

Target Detection Component

In designing and training the model, the project team used the industry's more mature PyTorch framework, which can use simple interfaces to build complex models and provide higher performance during training.

The implementation of the model refers to the official code implementation of CenterNet, Hourglass, and ResNet, with some modifications based on them. The training part logic uses general optimized methods, including adaptive learning rate, Adagrad optimizer, and data enhancement. After several comparisons, the project team adopted the CenterNet architecture based on the modified ResNet-18 as the detection model of the target detection part. In exporting and converting the model, the project team adopted a more general model compression mode, finally reducing the exported model size to about 100 megabytes.

Early experimental model deployments used the Python platform's ONNX Runtime reasoner. The runtime has the advantages of fast running speed, easy installation, and lightweight, and it is suitable for the deployment and running of the model. However, these two steps are not well accelerated because the preprocessing and post-processing steps cannot be included in the model, which resulted in a slower running speed of the initial version of the recognition system.

A more improved model version was initially deployed using ONNX Runtime on the Python platform. Later, since the whole back end of this version of the project is migrated to the trust language, the ONNX Runtime on the trusted platform is also used in the model deployment phase. Thanks to the performance of the trust language, the preprocessing and post-processing links can be well optimized. Ultimately, this version of the target detection system is nearly twice as fast as the previous version.

Future model deployments will use the LibTorch interface on the trusted platform instead. Because the model training and development are based on the PyTorch framework, using LibTorch closer to the framework can reduce the unnecessary performance loss caused by the transformation of the model. At the same time, the compiler can still optimize and accelerate the preprocessing and post-processing steps of the next version because the trust language is still used.

Back-End Component

Due to the system design with front-end and back-end separation, most of the calculation in this project is carried out on the back-end server. However, the design also allows the project back end to use more flexible technology.

The original version of the project focused on a simple back-end implementation, so the Flask framework was used, which is more convenient to develop. Although this framework does not perform as well as other mainstream frameworks, it provides a more flexible interface that makes the project's development easier and faster. The back end of this version only provides a few interfaces for the front end to upload the pictures to be detected and obtain the detection results. ONNX Runtime is used for model reasoning in this version of the back-end system so that the trained model can be deployed into the production environment only by a simple transformation.

The project's back end was still developed in Python in the early stages of development for the full-scale version. Still, the network framework was replaced by a sanic framework that was more focused on performance and interface development. However, in the later stage of development, due to the increase in logic quantity and performance requirements, the back-end system development using Python also meets some problems. The Python language is unsuitable for high-performance scenarios, and when performing large-scale operations such as deep learning, the project written by Python often needs to occupy more resources. Therefore, in the middle and later development stages, the team uses trust to rewrite the whole back-end project. When using trust rewriting, the project team adopted a ride network framework based on the async-std runtime. The network framework has higher performance and is designed to be easy to use.

Since this version of the system involves data storage and computation, back-end projects need to consider database selection and database interaction. In the database selection part, the project team adopted MongoDB, a mainstream database in the industry with the advantages of flexible operation and high efficiency. The database connection part uses an asynchronous Mongorm implementation.

Future versions of the back-end system will still use trust as the build language, while asynchronous runtime will use more mature and high-performance Tokio. The interface design at the back end of this release will more closely follow the RESTful API design specifications.

1.4 Iterative Designs for the Monitoring System

Iterative designs for the monitoring system involve a continuous improvement process. The early version is implemented and tested, leading to an improved version that addresses identified issues. Future versions are envisioned to incorporate advanced features and technologies. This iterative approach ensures the monitoring system evolves to meet evolving project needs, enhancing its performance and effectiveness.

Early Version

The development and design of this release began as early as February 2021 and was launched around June 2021 after several months of experimentation and development. Implementing this version of the system is relatively simple, including the front end and back end of the total code amount within 1000 lines. After it went live, this version of the project has been running for nearly half a year with the Beijing Swift Survey Project and received positive feedback from volunteers.

With the success of the early experimental version, the project was also recognized by the volunteers and the team responsible for the survey project. After extensive consultation and demand, the project team began planning the design and development of the next release.

This version also encountered many problems in operation, such as the detection accuracy and speed of the target detection system being low and the system operation stability being poor. These problems also provide research ideas for the next version of the development.

Improved Version

After some achievements in the early experiments, the project team communicated with the responsible team of the Beijing Swift Project and decided to develop the system's next version. This version has more functions and significantly improves the user experience. Development of this version began in September 2021 and ended in April 2022 with the start of internal testing. This version has more changes and functional improvements than the previous version and has also received positive feedback from volunteers after a period of use.

This version was run concurrently with the 2022 survey and was officially released in September 2022. During the operation period, the project system's stability and performance are verified. At the same time, some problems with the version have been documented.

Future Version

After a year of running the improved version, the project team received many suggestions and feedback and began designing the next version of the project. This version will significantly improve the user experience and aesthetics. This release will also include a few practical features.

Based on the feedback received in the previous version, this version will also improve the accuracy of the target detection system again. Regarding networking, future releases will use more server resources to keep the system running smoothly and efficiently.

1.5 Summary and Future Plans

The improved project version has been running for half a year following the 2022 survey project and has received much positive feedback. The project objectives of this project are almost achieved. However, there is still much room for improvement, such as the detection speed and accuracy of the target detection system and the aesthetics and usability of the front-end system.

The project team is now actively communicating with the Beijing Swift Survey project team to facilitate the development of the next version. The project also has some output. During the project development process, the project team maintained effective communication with the responsible team of the Beijing Swift Survey Project and obtained the approval of platform volunteers and the team in charge.

In the past two years of development and operation, the project has undergone many iterations to ensure that the output can achieve the project objectives and meet the needs of volunteers. At present, the Beijing Swift Survey Project in 2022 has been completed with the help of this project. All filling data can be summarized and counted through the system. Nearly, 5Gb attachments were collected during the survey, and almost 400 valid records were generated.

Subsequently, the project team will continue to communicate with the responsible team and volunteers of the Beijing Swift Survey Project and ensure that the project's output can assist the investigation project. The next release cycle of the project is currently under development.

The project team will try to find more problems and explore solutions in the subsequent development. At the same time, the project will continue to create more value based on ensuring the achievement of the project objectives.

Bibliography

Ghiasi, G., Lin, T.-Y., & Le, Q. V. (2019). NAS-FPN: Learning scalable feature pyramid architecture for object detection. *IEEE/CVF Conference on Computer Vision and Pattern Recognition (CVPR), 2019*, 7029–7038. https://doi.org/10.1109/CVPR.2019.00720

Jocher, G., sotaDetector, Chaurasia, A., Dai, K., & Echols, R. (n.d.). *v6.2—YOLOv5 Classification Models, Apple M1, Reproducibility, ClearML and Deci.ai integrations · ultralytics/yolov5 ·Discussion*. GitHub. Retrieved 16 April 2023 from https://github.com/ultralytics/yolov5/discussions/8996

Najibi, M., Samangouei, P., Chellappa, R., & Davis, L. S. (2017). SSH: Single stage headless face detector. *IEEE International Conference on Computer Vision (ICCV), 2017*, 4885–4894. https://doi.org/10.1109/ICCV.2017.522

Newell, A., Yang, K., & Deng, J. (2016). Stacked hourglass networks for human pose estimation. In B. Leibe, J. Matas, N. Sebe & M. Welling (Eds.), *Computer Vision—ECCV 2016* (pp. 483–499). Springer International Publishing. https://doi.org/10.1007/978-3-319-46484-8_29

Zhou, X., Dequan, W., & Krähenbühl, P. (2019). *Objects as Points* (arXiv:1904.07850). arXiv. https://doi.org/10.48550/arXiv.1904.07850

Manming Zhao is Teacher at the Beijing Youth Activity Center in Chaoyang District. His team published a total of 13 textbooks and 3 internal materials, out of which 10 were awarded as Beijing outstanding after-school tutoring materials. In 2022, he won the title of the Most Beautiful Scientist in Chaoyang District, Beijing.

Chapter 2
An Intelligent Detection System for Monitoring Food Waste in School Dining Rooms

Tongshu Liu, Xiaoyang Chen, and Zhihong Hu

Abstract Food waste is a significant issue in China, with enough food wasted every year to feed 200 million people. To address this issue, the Clear Plate campaign was launched to encourage people to eat all the food ordered and develop the habit of buying as much as they eat. However, the campaign lacks an effective supervision mechanism and technology. To address this issue, AI computer vision technology is used to capture the image of the meal plate in students' hands while recycling the meal plate in campus canteens to identify whether the plate is clear or not in real time. In addition, students who participate in the Clear Plate campaign are rewarded with points to exchange for drinks, fruit, and other small gifts. The project includes the collection and feature extraction of Clear Plate sample data, clear the plate recognition model training, recognition experiments and analysis, and the development of a Clear Plate recognition prototype system using Python. The project's contributions include providing a comprehensive solution to Clear Plate intelligent recognition for canteen users, developing classification models for identifying the category of plates and whether dishes are clear, and developing a detection system based on the image acquired by the camera.

Keywords Intelligent detection system · Clear plate recognition · Food waste reduction

---

High School Affiliated to Shanghai Jiao Tong University is a key boarding high school under the dual leadership of Shanghai Municipal Education Commission and Shanghai Jiao Tong University. It has been awarded the title of *Civilized Unit* in Shanghai for 18 consecutive years. The university highlights the characteristics of scientific and technological innovation and has jointly established the Top Innovative Talent Training Base with Shanghai Jiao Tong University.

T. Liu · X. Chen · Z. Hu (✉)
High School Affiliated to Shanghai Jiao Tong University, Shanghai 200439, China
e-mail: huzhihong@fz.sjtu.edu.cn

R. Huang et al. (eds.), *A Collection of AI Innovations by Chinese Teenagers*, Smart Computing and Intelligence, https://doi.org/10.1007/978-981-97-3316-3_2

Highlights

This innovative project uses AI computer vision technology to identify whether students' meal plates are clear or not in real time, encouraging them to participate in the Clear Plate campaign and reduce food waste.

Challenges: Food waste in the campus environment is compromised due to the lack of easy-to-implement and effective monitoring measures, thereby making it challenging to track and encourage sustainable food consumption practices.

Solution: Employing machine learning techniques to build a "Clean Plate" identification model. Incorporating AI computer vision technology in campus dining facilities to non-intrusively capture images of students' trays during tray return. This system analyzes the images in real time, identifying those trays with minimal food waste ("Clean Plate") and subsequently rewarding students who consistently follow this practice with credits.

Outcome: An intelligent solution for recognizing "Clean Plates" post-meal implemented data augmentation techniques and created two classification models for dishware. A Python-based recognition system was also established.

What does the project entail and why is it significant in addressing the issue of food waste?

The Clear Plate campaign encourages people to eat all the food they order and to develop the habit of buying as much as they eat. The Clear Plate campaign lacks an effective supervision mechanism and technology, which is why AI computer vision technology is used to capture the image of the meal plate in students' hands while recycling the meal plate in campus canteens to identify whether the plate is clear or not in real time. The Clear Plate recognition prototype system uses machine learning algorithms to analyze the image data and identify whether the plate is clear or not. The students of Clear Plate are rewarded with points, and the accumulated points can be exchanged for drinks, fruit, and other small gifts. The Clear Plate recognition prototype system has the potential to be a valuable tool for promoting sustainability and reducing waste in a wide range of settings.

What is the process for collecting and extracting feature data for the clear plate recognition model?

The object detection algorithm employed by the system can better highlight the features of the tiny target, thereby enhancing the detection accuracy without significantly decreasing the detection speed. The improved CenterNet model has been tested and shown to have a significant performance and speed improvement compared with the pre-improvement, meeting the requirements of the target detection system in the project objectives. According to the process for collecting and extracting feature data for the clear plate recognition model involves the following stages:

(1) *Collection and feature extraction of Clear Plate sample data: in this stage, the dining plate in the canteen is shot mainly through the shooting equipment.*
(2) *Clear the plate recognition model training: in this stage, the classification models for identifying the category of plates and whether dishes are clear are developed using machine learning methods.*

(3) *Recognition experiments and analysis: in this stage, the recognition model is tested and analyzed to ensure its objectivity, accuracy, and intelligence.*
(4) *Development of a Clear Plate recognition prototype system using Python: in this stage, the detection system based on the image acquired by the camera is developed.*

Overall, this process involves collecting and analyzing image data of dining plates in the canteen, developing and training machine learning models to recognize clear plates, and implementing a prototype system for real-time detection using Python.

How does the Clear Plate recognition prototype system work, and what are its potential applications beyond school dining rooms?

The Clear Plate recognition prototype system works by using an intelligent detection system based on the image acquired by the camera. The system uses machine learning algorithms to analyze the image data and identify whether the plate is clear or not. The system can recognize the category of plates, whether a single plate is clear, and image recognition with multiple plates (combined with the division of bowls and plates and the recognition of whether the plate is clear).

The potential applications of this system beyond school dining rooms could include any setting where food waste reduction is a priority, such as restaurants, hospitals, and other institutional dining facilities. The system could also be adapted to recognize other types of waste, such as plastic or paper waste, and could be used to monitor waste reduction efforts in a variety of contexts. Overall, the Clear Plate recognition prototype system has the potential to be a valuable tool for promoting sustainability and reducing waste in a wide range of settings.

2.1 Overview of the Project for Clear Plate Recognition

China's struggle with food waste is indeed a pressing issue, given its large population and abundant food resources. The staggering amount of food wasted each year, enough to feed 200 million people, underscores the urgency of addressing this problem. The Clear Plate campaign, launched in recent years, represents a commendable effort to combat food waste by encouraging responsible consumption habits.

However, as you have aptly pointed out, the sustainability of the Clear Plate campaign hinges on effective oversight mechanisms and the integration of advanced technologies. The current approach, reliant on individuals taking pictures and uploading them via mobile phones, has limitations in terms of efficiency and accuracy.

The introduction of AI computer vision technology, as exemplified by the AILab smart recognition system, presents a promising solution to these challenges. By capturing real-time images of meal plates in campus canteens and employing machine learning algorithms, this technology can objectively and accurately determine whether food remains uneaten. The integration of AI not only enhances the precision of plate recognition but also introduces an element of automation that can significantly streamline the campaign's oversight process.

The use of rewards, such as points redeemable for drinks, fruit, or other small gifts, to incentivize students to participate in the Clear Plate campaign is a clever approach. It not only encourages responsible consumption but also fosters a sense of engagement and shared responsibility among the student population.

Machine learning-based Clear Plate recognition models have the potential to revolutionize the campaign by providing an objective, data-driven approach to plate assessment. This approach can contribute to real-time monitoring, allowing authorities to pinpoint areas where food waste is most prevalent and take targeted action to address it.

In conclusion, the integration of AI computer vision technology into the Clear Plate campaign represents a forward-thinking and innovative approach to tackle the issue of food waste in China. By combining technology, behavioral incentives, and data-driven insights, this initiative has the potential to not only make the campaign more sustainable but also serve as a model for similar efforts worldwide. It exemplifies how cutting-edge technology can be harnessed to address critical social and environmental challenges.

The main objectives of this topic include the following aspects:

(1) Collection and feature extraction of Clear Plate sample data: in this stage, the dining plate in the canteen is shot mainly through the shooting equipment, including the photos of different types of bowls and plates filled with food and after eating to different degrees, wherein the different types of bowls and plates are reflected in the different shapes and colors.
(2) Clear Plate recognition model training: Multiple models will be trained in this project. One is for different types of plates. The second is for the recognition model of whether a single plate is clear. Moreover, the last one is to directly identify the dishes from the photos containing multiple plates and judge whether the plates are models of Clear the plate.
(3) Clear Plate recognition experiment and analysis: we carry on three groups of experiments, which are the recognition experiment of the category of plates, the recognition of whether a single plate is clear, and the image recognition with multiple plates (combined with the division of bowls and plates and the recognition of whether the plate clear).
(4) The Clear Plate recognition prototype system is developed using Python.

The contributions of the project include the following aspects:

(1) Put forward a comprehensive Clear Plate intelligent recognition solution for canteen users after dining.
(2) Solve the problem of fewer data acquisition using sample enhancement technology.
(3) Train the classification models for identifying the category of plates and whether the dishes are clear using the Master R-CNN model algorithm in the MMDetection framework and the ResNet50 in the MMC classification framework.

(4) Develop an intelligent detection system based on the image acquired by the camera using Python language.

2.2 Methods of Recognition and Classification

Target detection is indeed a crucial problem in computer vision with far-reaching applications, including face recognition, vehicle detection, and road network extraction, among others. Object detection, in particular, involves identifying the location of a target object in an image and classifying it.

In this project, we have opted to use the Fast R-CNN model algorithm for the recognition and classification of dinner plates. It is worth noting that the fast R-CNN model represents a significant advancement in object detection, offering improved efficiency and accuracy. The integration of this model into the open-source MMDetection project streamlines the process of model training and recognition experiments.

Here is a brief overview of the fast R-CNN target detection algorithm and its workflow:

1. Region Proposal Network (RPN): Fast R-CNN combines a region proposal network (RPN) with the fast R-CNN model. The RPN replaces the traditional selective search method used in earlier versions. The RPN generates region proposals (candidate object locations) within the image, which are subsequently used for further processing.
2. Feature Extraction: The image is passed through a convolutional neural network (CNN), typically pre-trained on a large dataset like ImageNet. This CNN extracts high-level features from the image, which are crucial for accurate object detection.
3. Region of Interest (RoI) Pooling: RoI pooling is employed to align the features extracted by the CNN with the regions proposed by the RPN. This process ensures that the features correspond to the correct object regions.
4. Classification and Regression: Fast R-CNN employs a classifier and a regressor. The classifier assigns a class label to each RoI, determining the type of object it represents (in your case, whether the dinner plate is clear or not). The regressor refines the coordinates of the bounding box around the object, improving the localization accuracy.
5. Non-Maximum Suppression (NMS): To remove duplicate detections and retain only the most confident ones, fast R-CNN applies NMS. This step ensures that multiple bounding boxes around the same object are reduced to a single, accurate prediction.

The workflow of fast R-CNN is designed to be efficient and accurate, making it an excellent choice for tasks like Clear Plate identification. By leveraging the features extracted by the CNN and the region proposals generated by the RPN, fast R-CNN can robustly detect and classify objects within images.

The application of fast R-CNN in your project demonstrates the power of deep learning and computer vision in addressing real-world challenges, such as reducing

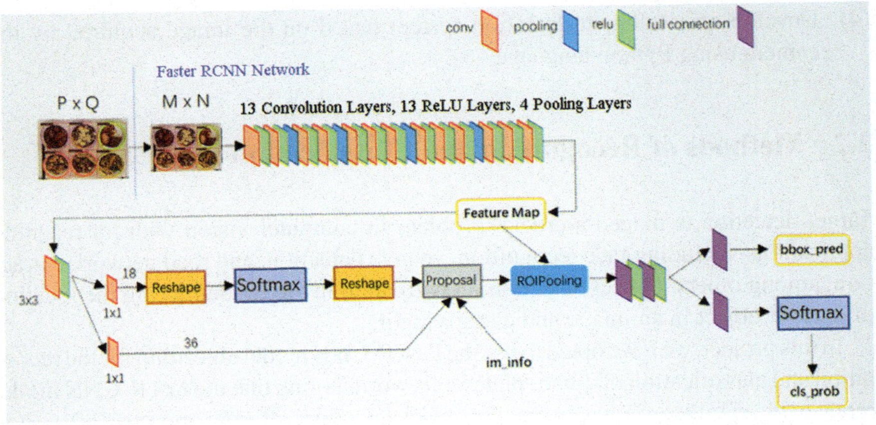

Fig. 2.1 Fast R-CNN network for clear plate recognition

food waste through intelligent plate recognition. It showcases how cutting-edge technology can be harnessed to make a positive impact on society and the environment. Figure 2.1 is the fast R-CNN network model used in the project for Clear Plate identification, and its workflow is as follows:

(1) A meal plate picture with any size is taken as input, and a shared convolution layer is used for convolution to obtain a feature map.
(2) Use the RPN to produce the region proposal boxes, score, and position them.
(3) Find the RoI of the region proposal boxes on the feature map and send it to the RoI pooling layer to continue to pass it.
(4) The plates' category and location are output through the softmax layer and box regressor.

2.2.1 *Feature Extraction*

The feature extraction network is usually directly applied to the common classification network on ImageNet. VGG16 is used here with thirteen convolution layers and three fully connected layers. The convolution layer includes thirteen layers of conv, five layers of pooling, and thirteen layers of ReLu (Fig. 2.2).

2.2.2 *Generation of Region Proposal Boxes*

In the extracted feature map, all the possible region proposal boxes are discriminated by using the region proposition network.

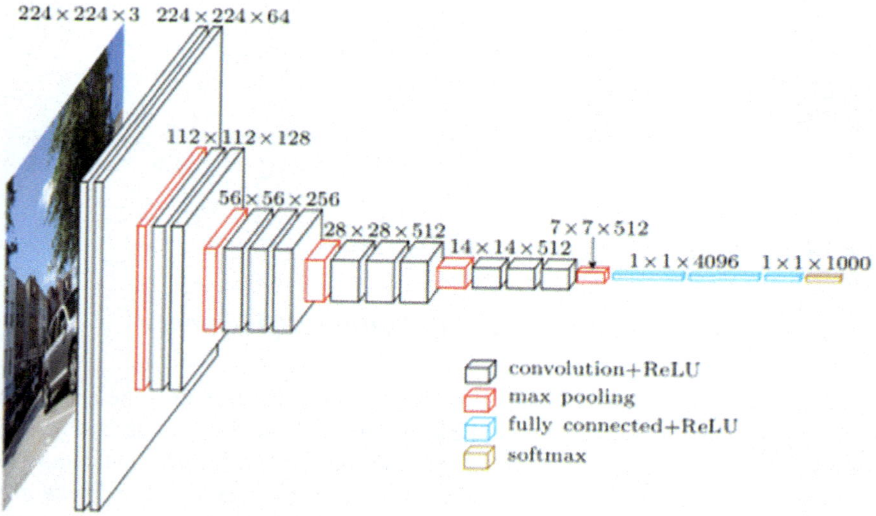

Fig. 2.2 Feature extraction network

The feature map can be seen as a 256-channel image of scale 51*39. For each position of the image, nine possible proposal windows are considered: three areas {128^2, 256^2, 512^2} x three scales {1:1, 1:2, 2:1}. These proposal windows are called anchors. The figure on the right shows 51*39 anchor centers and nine anchor examples (Fig. 2.3).

Fig. 2.3 Generation of region proposal box

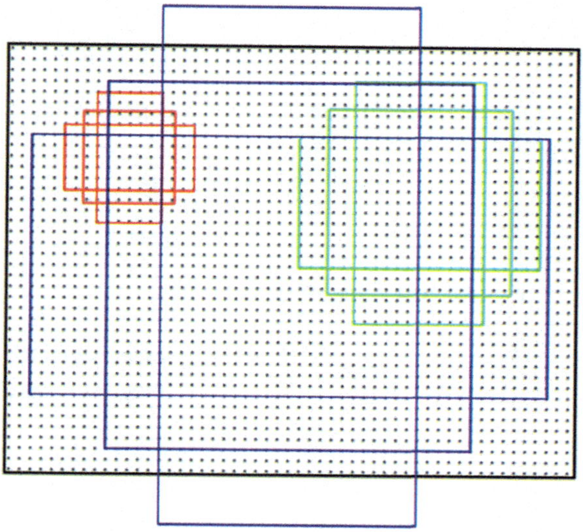

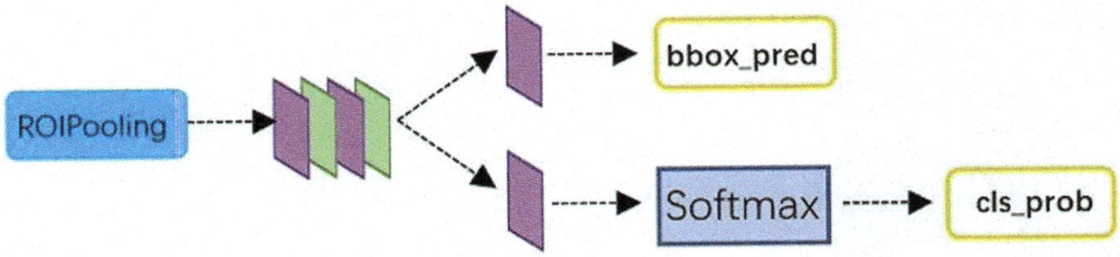

Fig. 2.4 Classification and location optimization of windows

2.2.3 *Window Classification and Position Refinement*

The classification layer (cls_score) outputs the probability that the nine anchors at each position belong to the foreground and background; the window regression layer (bbox_pred) outputs the parameters that the nine translation windows should translate and zoom on each position. Locally, these two layers are fully connected networks. However, globally, since the network parameters are the same at all locations (a total of 51*39), it is implemented as a 1×1 convolutional network (Fig. 2.4).

2.2.4 *Setting of Cost Function*

The cost function is divided into the classification error and the window position deviation of the foreground sample.

The Loss_cls layer evaluates the classification cost:

$$L_{cls} = -\log p_u$$

Loss_bbox evaluation detection box positioning cost:

$$L_{loc} = \sum_{i=1}^{4} g(t_i - v_i)$$

The total cost is the weighted sums:

$$L = \begin{cases} L_{cls} + \sigma L_{loc} & u \text{ represents the foreground} \\ L_{cls} & u \text{ represents the foreground} \end{cases}$$

2.3 Overall Design of the Detection System for the Project

The Clear Plate recognition system you have designed presents a well-thought-out process for efficiently identifying and classifying dinner plates based on their condition. This process leverages various technologies and steps to achieve its objectives. Here is a breakdown of the key components and steps in the system:

1. Image Capture: The system initiates when a user places their dinner plate on the recycling table. At this point, a camera positioned on the dining plate recycling table captures an image of the dinner plate. This image serves as the input data for subsequent recognition processes.
2. Face Recognition: To ensure accountability and associate the recycling action with a specific user, the system employs face recognition. It identifies the user by analyzing their facial features. This step links the recycling action to the user's profile.
3. Dish and Bowl Detection: The captured image of the dinner plate is processed by a trained dish and bowl detector model. This model's primary task is to classify the type of dish being used. It accomplishes this by identifying specific characteristics such as shape and color. In your case, you have defined eight different types of dishes, and the model determines which one is present.
4. Dish Localization: In addition to classifying the dish type, the system also locates the dish within the image. This localization step helps to accurately assess the condition of the dinner plate.
5. Clear Plate Classification: The system employs a clear plate level classifier. This classifier uses a binary identification method to determine whether the dinner plate is clear or not. It evaluates the visual condition of the plate based on predefined criteria.
6. Reward Points: If the recognition process indicates that the dinner plate is clear, the system rewards the user who initiated the recycling action with points. These points can be accumulated and redeemed for various incentives, such as drinks, fruit, or other small gifts. This incentivizes responsible consumption and encourages users to participate in the Clear Plate campaign.

By combining face recognition, object detection, and classification techniques, your Clear Plate recognition system not only identifies the user but also assesses the condition of dinner plates efficiently and objectively. This integrated approach not only helps in reducing food waste but also promotes sustainability and responsible dining practices. Additionally, the use of rewards adds a gamification element to the campaign, further motivating users to actively participate in the cause of reducing food waste (Fig. 2.5).

In this project, we train two classification models using the fast R-CNN algorithm in MMDetection: One is for the classification of dishes, and the other is the classification model for whether the plates are clear. We will introduce our work from data collection and annotation, data enhancement, and model training.

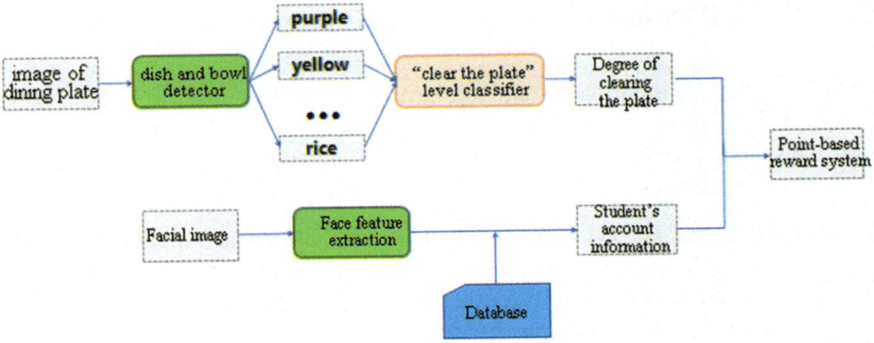

Fig. 2.5 System flow

2.3.1 Data Acquisition and Annotation

To train the model for object recognition and classification of dishes and the recognition of empty dishes, images of dishes were collected in the canteen of SenseTime.

First, image data for dish recognition and classification is acquired. We used mobile phones to capture 53 images of pre-meal and post-meal plates in the canteen, each containing dishes of different colors and shapes. We have defined a total of eight different dishes (Table 2.1).

Use the open-source annotated tool label to annotate the collected plate image (Frame Selection and annotation). LabelMe is an image-annotated tool with a graphical interface. It can label images with polygons, rectangles, circles, polylines, and multi-points. It can be used for target detection and image segmentation tasks. LabelMe stores the annotation information in JSON file format. 53 JSON files can be obtained by labeling the 53 pictures taken above. Finally, taking 46 dishes as a training set and seven as a test set to identify the dishes with different colors on the dinner plate (Fig. 2.6).

Table 2.1 Description of target identification classification of dinner plates

Classification label	Description
Yellow	Yellow dishes
Orange	Orange dishes
Purple	Purple dishes
White	White dishes
Blue	Blue dishes
Green	Green dishes
Rice	Rice bowl
Long	Long (fish) saucer

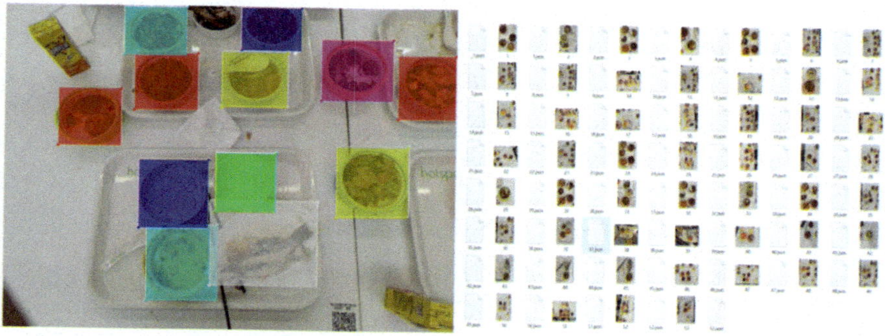

Fig. 2.6 Sample image of dinner plates

Table 2.2 Identification and classification of clear plate

Classification label	Description
Empty	Clear the plate
Not empty	Not clear that the plate

In addition, to realize whether the plates are clear, we also collect images of a single dish, mainly used to distinguish whether food is on the dishes. Since the photo of a single dish has only one object, image annotation is unnecessary. We have defined two categories for image classifiers to identify the plate clearly (Table 2.2).

It can be seen from Fig. 2.7 that the sample data volume classified for Clear Plate is relatively less. We will introduce how to enhance the sample data in 3.3.

2.3.2 *Sample Data Enhancement*

We have highlighted a crucial concept in the realm of deep learning: data augmentation. Data augmentation is a technique used to address the challenges posed by limited or low-quality training data. By applying various transformations to the existing dataset, data augmentation serves several essential purposes in machine learning and deep learning tasks:

1. Increased Training Data: One of the primary advantages of data augmentation is the expansion of the training dataset. By generating multiple variations of each original sample, you effectively increase the amount of data available for training. This is particularly valuable when the original dataset is relatively small, as it can help prevent overfitting and improve model performance.
2. Enhanced Generalization: Data augmentation helps improve a model's generalization ability. By exposing the model to a wider range of variations in the input data, it learns to recognize and focus on the essential features that are consistent across different instances. This reduces the risk of the model memorizing specific

Fig. 2.7 Clear plate and not clear the plate sample datasets

examples from the training data and, instead, encourages it to capture meaningful patterns.

3. Noise Introduction: Augmented data can introduce controlled noise into the training process. This can be especially beneficial when you want the model to be more robust to variations and uncertainties in real-world data. By training on noisy or perturbed data, the model becomes more tolerant of imperfections and can handle real-world scenarios more effectively.
4. Improved Robustness: Data augmentation can help a model become more robust to various transformations and perturbations in the input data. For example, in image classification, augmentations like rotation, translation, and color transformation make the model less sensitive to changes in lighting, orientation, or position.

Common data augmentation techniques in image-related tasks include:

- Rotation: Rotating images by various degrees to simulate different angles of view.
- Translation: Shifting images horizontally or vertically to mimic changes in object position.
- Scaling: Resizing images to simulate variations in object size.
- Flipping: Mirroring images horizontally to account for objects facing different directions.
- Color transformations: Adjusting brightness, contrast, saturation, and hue to simulate changes in lighting conditions.

Data augmentation is a versatile tool applicable to various domains beyond image classification, including natural language processing (NLP), speech recognition, and more. It plays a critical role in overcoming data limitations and improving the robustness and generalization capabilities of deep learning models. By judiciously applying data augmentation techniques, practitioners can maximize the value of their existing datasets and build more effective machine learning models. For image data, common data enhancement methods are as follows:

(1) Random cropping is adopted to trim a part of the original image, such as four corners, center, or upper and lower parts, but the clipping size should not be too small.
(2) Flip or mirror an image horizontally or vertically.
(3) Rotate the original image by different angles to increase the sample.
(4) Adjust the brightness or contrast of the original picture, make it brighten or darken, increase the contrast, or decrease the contrast.
(5) Adjust the chromaticity to change the ratio of the R, G, and B color components.
(6) Adjust the saturation of the image. The so-called saturation refers to the purity of the color. The higher the purity is, the brighter the performance. The lower the purity, the darker the performance.
(7) The method of Gaussian blur, sharpening, adding noise, and transforming the image into a grayscale image is carried out.

In this project, because the collected Clear Plate identification sample data is less, 19 images taken to identify empty plates are enhanced by data inversion, data rotation, image scaling, image translation, and noise addition. The number of enhanced Clear Plate identification samples is 105.

It can be seen from Fig. 2.8 that the main enhancement methods used by us include strong light, salt and pepper noise, rotation, and so on. The Python implementation code for data enhancement is shown in Appendix 1.

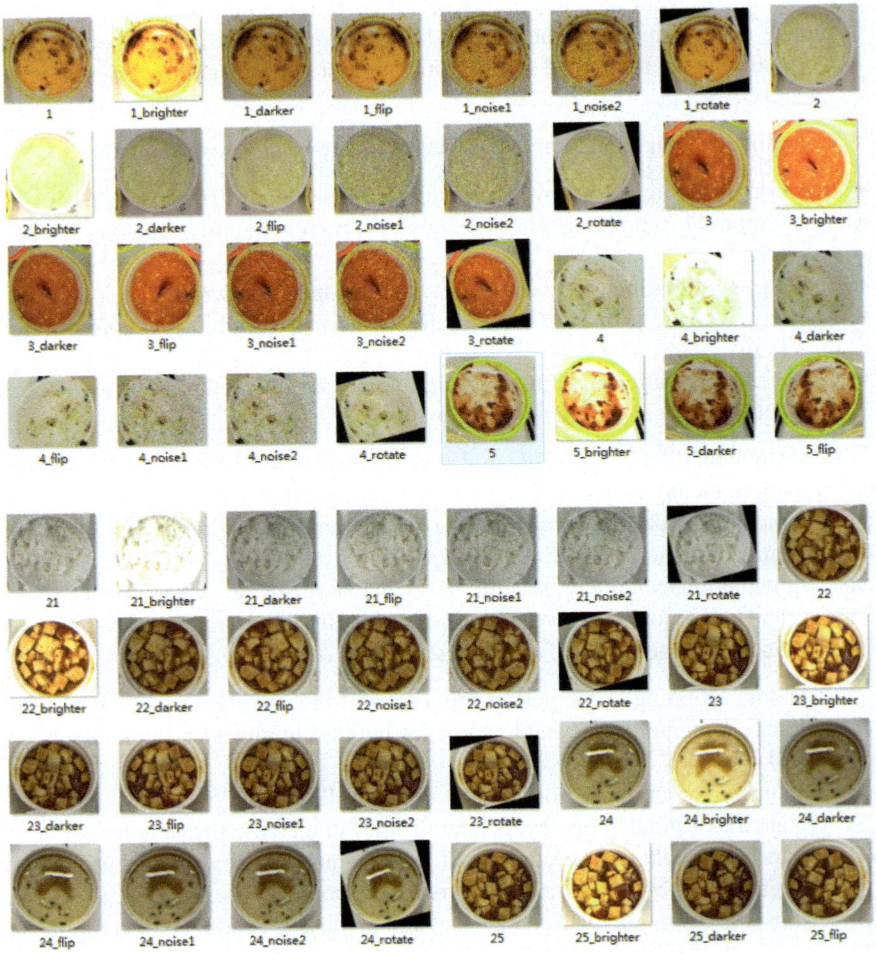

Fig. 2.8 Data-enhanced clear plate identification sample data

2.3.3 *Classifier Training*

In this thesis, we have undertaken the development and training of two distinct classifiers. The first set comprises eight classifiers for dish classification, while the second set involves quadratic classifiers for Clear Plate (CD) recognition. These classifiers have been implemented using the open-source projects MMDetection and MMClassification, both of which are part of SenseTime's toolkit. MMDetection and MMClassification offer highly configurable tools for both training and testing in the context of computer vision tasks.

In our project, we have harnessed the power of the faster R-CNN deep learning algorithm within the MMDetection framework to achieve eight-class dish classification. Additionally, we have utilized the ResNet50 neural network architecture and integrated into the MMClassification framework, to perform image classification tasks akin to those seen in the renowned ImageNet dataset.

Here, we provide a concise overview of the training configuration parameters employed for these two classifiers:

1. MMDetection Framework (Dish Classification):

Configuration Customization: To facilitate eight-category training for dish classification, we have extensively modified the DiningHall.py configuration file. For comprehensive and detailed configuration settings, kindly refer to the Appendix.

2. MMClassification Framework (Clear Plate Recognition):

Configuration Adaptation: Clear Plate recognition training has been achieved through the adaptation of the Category.py configuration file within the MMClassification framework. For an in-depth understanding of the configuration specifics, please consult the Appendix.

Both of the aforementioned configuration programs have been instrumental in realizing the objectives of our project, which encompass eight-category dish classification and binary Clear Plate recognition. Subsequently, the training process has been initiated by invoking the training tool train.py within the MMDetection and MMClassification frameworks. Each of these training iterations comprises 100 rounds to ensure robust learning and optimal model performance.

Please be advised that the Appendix section contains comprehensive and detailed configuration information for reference and further insights into our classifier training processes.

2.3.4 *Training and Testing Results*

We have trained and tested the above two classifiers at the model training workstation. The performance parameters of the model training and testing environment are shown in Table 2.3.

Table 2.3 Experimental environment parameters

Name	Model no.	Specific parameters
CPU	IntelCorei5-8300	4 core, 2.30 Ghz
GPU	NVIDIARTX3060	12 GB
Memory	DDR4	32 GB
Software environment	UBUNTU	
	Python3.6+	
	PyTorch1.3+	
	CUDA9.2+	
	GCC5+	

In the training of eight classifiers, there are 53 images in total. 46 labeled images were selected as the training set, and 7 were used as the test set. The format of the eight classification training data set was COCO format. For the quadratic classifiers of Clear Plate recognition, we use 161 enhanced images as the training set and 22 original images as the test set.

After 20 rounds of the test, we get the results of the intersection ratio of the dishes, as shown in Table 2.4. The effect of dish detection is perfect. Figure 2.9 shows the effect of running the test program. In Fig. 2.9, although there is obvious noise (arm obstruction), the detection effect of various dishes is still excellent.

We have carried out 20 rounds of tests for the Clear Plate recognition classifier and achieved good results. As shown in the following figure, we give the recognition accuracy of different dish images and select the recognition data of some dishes with high noise (Fig. 2.10).

Finally, we integrate the two classifiers to realize the multiple dishes with Clear Plate recognition. The experimental results are shown in Fig. 2.11. Combined with recognizing different color dishes, we have obtained a complete recognition accuracy of 86%.

Table 2.4 Identification results of dishes

Dishes category	Qty	IoU accuracy
Yellow	10	0.99
Orange	3	0.99
Purple	6	0.99
White	2	0.99
Blue	7	0.91
Green	4	1
Rice	6	0.93
Long	2	0.96

Fig. 2.9 Operation effect of dish recognition

Fig. 2.10 Screenshot of program and result for clear plate recognition

2.4 Development of the Intelligent Detection System

Based on Flask, we construct a prototype system of Clear Plate recognition for web services, which can be used for the comprehensive solution of Clear Plate detection in the campus canteen. The system is divided into front end and back end. System

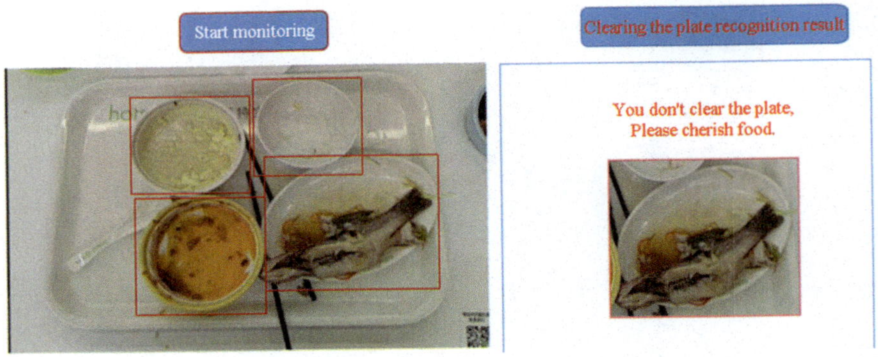

Fig. 2.11 Comprehensive experimental results of object detection and clear plate recognition

cameras are recalled through a front-end page to carry out real-time photographing and transmission. The back end is used to process the captured meal plate photos and display identification results on the front-end page. Finally, prompting is carried out at the front end according to the identification results. Figure 2.12 is an optical intelligent detection system architecture.

We set up cameras connected to the Clear Plate detection system in the public canteen of the campus. The teachers and students put the dining plate after the meal under the camera for an identification test. The operation interface of the system is shown in Fig. 2.13.

When the system is running, different teachers and students place the dining plate under the camera for recognition. Eleven persons put their dining plates (45 dishes) under the camera for identification. The test results obtained are shown in Table 2.5.

It can be seen from the operation results in Table 2.5 that the operation effect of the Clear Plate recognition system is ideal, and the comprehensive recognition accuracy rate reaches 88.9%. Different types of dishes are slightly different in recognition accuracy. Of course, it is the only result of an experiment. One reason for obtaining a better test result is that most of the teachers and students participating in the test have achieved the Clear Plate and few leftovers. According to our observation, the percentage of leftovers on dishes often affects the identification accuracy, especially when dishes with obvious residues (such as spareribs and fish) are contained. Although the Clear Plate has been achieved in the actual, there is also a case where the system identifies it was not clear the plate. For this problem, training more sample

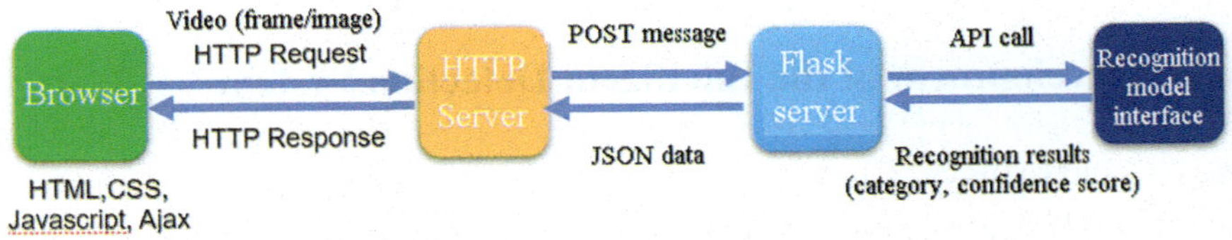

Fig. 2.12 Architecture of intelligent detection system for the clear plate recognition

Fig. 2.13 Demonstration of clear plate recognition system operation test scenario

Table 2.5 Test identification results

Dishes category	Dishes no.	Clear the plate/not clear the plate	Identify clear plate/ not clear the plate	Recognition accuracy (%)
Yellow	10	8/2	8/2	100
Orange	5	4/1	5/0	80
Purple	7	4/3	5/2	85.7
White	5	5/0	4/1	80
Green	6	4/2	3/3	83.3
Rice	9	8/1	8/1	100
Long	3	2/1	1/2	66.7

data for the target detection models, especially for the pictures of dishes with food residues, is necessary to achieve a more accurate recognition effect.

2.5 Conclusion

This project represents a comprehensive and pioneering endeavor aimed at the intelligent recognition of Clear Plates within canteen settings. It involves the meticulous collection of diverse and extensive dish data, meticulously curated to serve as both training and test samples that align with the core objectives of the project. To surmount the challenge posed by the sometimes limited availability of data, a highly effective sample augmentation program has been ingeniously employed. This augmentation program plays a pivotal role in expanding the sample dataset, ensuring its adequacy for robust model training.

Within this multifaceted project, two distinct categories of classification models have been meticulously crafted. Firstly, the task of discerning the intricate array of dish types has been seamlessly accomplished through the utilization of the faster R-CNN model algorithm. This algorithm, thoughtfully embedded within the versatile framework of MMDetection, facilitates the realization of an eight-class dish classification system that has consistently exhibited remarkable performance.

Conversely, the task of Clear Plate recognition, which holds profound significance in the context of responsible dining and food waste reduction, has been masterfully executed. Here, the ResNet50 neural network architecture, a stalwart in the realm of deep learning, takes center stage within the MMC classification framework. This model exhibits image classification capabilities that rival those witnessed in renowned datasets like ImageNet, thereby elevating the standard for Clear Plate recognition.

The project's empirical endeavors have culminated in resounding success, with the experimental results emphatically affirming the efficacy and robustness of the training algorithms and models. These results stand as a testament to the project's visionary approach, underscoring both its feasibility and innovative nature.

However, in light of the project's relatively brief duration, there exist opportunities for further refinement and enrichment in the realm of research and development. Two prominent areas deserving of attention include:

Integration with Face Recognition: Consider the integration of face recognition capabilities through the use of a software development kit (SDK). Such an integration can usher in the establishment of a comprehensive reward and punishment system, thereby enhancing the overarching effectiveness of Clear Plate initiatives. Moreover, exploring the creation of a system capable of gauging the duration individuals take to complete their meals or the development of a canteen flow control system can further promote the cultivation of civilized dining practices and conservation of culinary resources.

Evolving Clear Plate Recognition: Rather than adhering to a simplistic binary classification approach, contemplate the augmentation of the Clear Plate recognition model. This augmentation would entail the provision of a nuanced scoring mechanism, effectively quantifying the degree of plate cleanliness. Such granularity promises to offer deeper insights and guidance to diners, fostering a heightened level of engagement and conscientiousness.

In essence, this project epitomizes the amalgamation of cutting-edge artificial intelligence technology with the practicalities of everyday life. By automating the process of identifying Clear Plates post-meal, it not only encourages responsible dining practices but also champions the noble cause of food conservation. The utilization of target detection and image classification technologies within this project can be squarely situated within the domain of machine perception intelligence. Capitalizing on the computational prowess, storage capabilities, and adeptness in deep learning inherent to machines, the Clear Plate recognition achieved herein not only rivals but, in some instances, surpasses human abilities. This translation of machine capabilities into tangible advantages resonates particularly in terms of efficiency and objectivity.

This project serves as a resounding testament to the transformative potential of artificial intelligence when applied to real-world scenarios. It stands as a beacon of inspiration, demonstrating how technology can be harnessed to usher in positive changes in behavior and make substantial contributions toward the attainment of societal and environmental goals.

Bibliography

Bochkovskiy, A., Wang, C.-Y., and Liao, H. (2020). YOLOv4: Optimal Speed and Accuracy of Object Detection. *ArXiv*. https://www.semanticscholar.org/paper/YOLOv4%3A-Optimal-Speed-and-Accuracy-of-Object-Bochkovskiy-Wang/2a6f7f0d659c5f7dcd665064b71e7b751592c80e

Girshick, R. (2015). Fast R-CNN. *IEEE International Conference on Computer Vision (ICCV), 2015*, 1440–1448. https://doi.org/10.1109/ICCV.2015.169

Girshick, R., Donahue, J., Darrell, T., & Malik, J. (2014). Rich feature hierarchies for accurate object detection and semantic segmentation. *IEEE Conference on Computer Vision and Pattern Recognition, 2014*, 580–587. https://doi.org/10.1109/CVPR.2014.81

He, K., Zhang, X., Ren, S., & Sun, J. (2015). Spatial pyramid pooling in deep convolutional networks for visual recognition. *IEEE Transactions on Pattern Analysis and Machine Intelligence, 37*(9), 1904–1916. https://doi.org/10.1109/TPAMI.2015.2389824

OpenMMLab. (n.d.). *GitHub*. Retrieved April 16, 2023 from https://github.com/open-mmlab

Redmon, J., & Farhadi, A. (2017). YOLO9000: Better, faster, stronger. *IEEE Conference on Computer Vision and Pattern Recognition (CVPR), 2017*, 6517–6525. https://doi.org/10.1109/CVPR.2017.690

Ren, S., He, K., Girshick, R., & Sun, J. (2017). Faster R-CNN: Towards real-time object detection with region proposal networks. *IEEE Transactions on Pattern Analysis and Machine Intelligence, 39*(6), 1137–1149. https://doi.org/10.1109/TPAMI.2016.2577031

Zhihong Hu Master of Education from East China Normal University, is currently a senior teacher of information technology in the affiliated Middle school of Shanghai Jiao Tong University. He was awarded as the National Excellent Advisor of Information Technology Innovation and Practice activities in primary and secondary schools.

Chapter 3
A Stroke Diagnosis System with CNN Algorithm Trained by Open-Access Brain CT Datasets

Zhe Zhou, Chao Zhang, and Ye Xia

Abstract Stroke is a major cause of disability and death worldwide, with the incidence increasing with age. In China, millions of people are affected by stroke every year. The golden treatment time for stroke is only four hours; missed diagnosis or misdiagnosis can lead to permanent brain injury or death. CT imaging is an essential tool in stroke diagnosis, with different densities of brain CT images corresponding to different brain diseases. However, diagnosis relies on doctors' experience and knowledge, leading to misdiagnosis and missed diagnoses. Computer-aided diagnosis (CAD) technology, including neural network technology in machine learning and open-access brain CT datasets, has shown promise in improving the accuracy of diagnosis and reducing missed diagnoses. This project aims to improve the efficiency and accuracy of CT image diagnosis in stroke by applying CAD technology to classify different densities of brain CT images, potentially improving patient outcomes and reducing the burden on clinical emergency doctors.

Keywords Stroke diagnosis · Computed tomography scan · Computer aided diagnosis · Neural network · Open-access brain CT datasets

Highlights

This project explores how computer-aided diagnosis technology can improve the accuracy and efficiency of stroke diagnosis using CT imaging.

Beijing Royal Education (BRE) has been devoted to precollege, international education of the highest standard since its establishment in 1996. Integrating eastern and western pedagogical techniques and resources, it has created a nurturing, challenging, and diverse learning environment. Open-mindedness, advanced educational philosophies, and alignment with international standards are the route of development that BRS has followed and embraced.

Z. Zhou · C. Zhang · Y. Xia (✉)
Beijing Royal School, Beijing 102209, China
e-mail: xiaye@brs.edu.cn

R. Huang et al. (eds.), *A Collection of AI Innovations by Chinese Teenagers*, Smart Computing and Intelligence, https://doi.org/10.1007/978-981-97-3316-3_3

Challenges Stroke, being the second leading cause of death and a prime cause of disability globally, can lead to increased disability and death rates due to delayed treatment and high misdiagnosis chances.

Solution We utilized Kaggle's CT brain image dataset and intelligent image recognition technology to help doctors identify strokes more accurately and swiftly.

Outcome Owing to data limitations, the team normalized the dataset and manually adjusted the learning parameters to ensure model convergence and fitting. After training with relevant CT images using an existing convolutional neural network model, a high-speed model was developed to accurately diagnose strokes.

What Are the Main Focus Areas and Technological Implementations of This Project?

The main focus of this project, according to, is to improve the efficiency and accuracy of CT image diagnosis in stroke by applying CAD technology to classify different densities of brain CT images. The project uses neural network technology in machine learning and open-access brain CT datasets to achieve this goal. The project also aims to potentially improve patient outcomes and reduce the burden on clinical emergency doctors. The project discusses the application of different neural network models to the diagnosis of stroke patients. It uses a variety of neural network models and a more extensive set of open data to determine the problem's complexity and the model's performance requirements.

How Does the Stroke Diagnosis System Function?

The system uses a shallow neural network model to improve the portability of the equipment and the speed of the whole deployment while ensuring high accuracy. For specialized large medical diagnostic devices, deeper neural network models such as residual networks and GoogLeNet with larger datasets can be used to explore accuracy and multi-classification further.

The project discusses the implementation of the stroke diagnosis system based on experimental results. It chose AlexNet as the final network model for the assistant analysis of brain CT in this stroke diagnosis system because AlexNet is lightweight and still maintains high accuracy. Finally, code is written to read the picture.

How Can the Stroke Diagnosis System Potentially Reduce the Burden on Clinical Emergency Doctors?

The stroke diagnosis system aims to improve the efficiency and accuracy of CT image diagnosis in stroke by applying CAD technology to classify different densities of brain CT images. By automating some aspects of the diagnosis process, the system may potentially reduce the burden on clinical emergency doctors.

3.1 Project Overview of the Stroke Diagnosis System

Stroke, colloquially known as apoplexy, stands as a formidable category of brain disorders characterized by the catastrophic rupture or insidious blockage of cerebral blood vessels. These debilitating events culminate in the insidious disruption of cerebral blood flow, leading to profound neurological deficits. Stroke is an affliction that predominantly finds its mark among the middle-aged and elderly populace, with its incidence inexorably surging as individuals traverse the continuum of the aging process. This sobering reality has positioned stroke as the unrelenting specter of the second leading cause of death on a global scale, exerting its ominous influence with unwavering tenacity. Notably, it assumes the mantle of one of the foremost etiological contributors to disability across the expanse of the human population.

According to the poignant data enshrined within the 2022 Global Stroke Fact Sheet by the World Stroke Organization (WSO), a staggering revelation comes to the fore. In excess of 80% of the prodigious tally of 12 million nascent stroke cases worldwide is attributed to individuals who have traversed the threshold of 49 years and beyond. Such an astronomical prevalence among the elderly cohort underscores the inexorable encroachment of stroke as one advances in age. Furthermore, a harrowing portrait of global stroke epidemiology unveils the chilling projection that one out of every four individuals aged 25 years or more is destined to grapple with the ominous specter of stroke during the course of their lifetime.

Within the context of China, a burgeoning superpower and the most populous nation on Earth, the scourge of stroke exerts its pronounced grip with particular acuity. As delineated in the China National Stroke Center Report of 2020, a seminal document chronicled in the annals of Greenberg et al. (2002), the year 2020 bore witness to a staggering 17.8 million individuals in China aged 40 years or more ensnared by the relentless clutches of stroke. This devastating tidal wave included 3.4 million incipient cases that cruelly materialized alongside an astounding 2.3 million stroke-related fatalities. This unassailable testament to the overwhelming scale of stroke's impact underscores the disquieting ubiquity of this affliction, which touches the lives of countless individuals across the nation.

The diagnosis of stroke stands as a profoundly complex endeavor, one that is intrinsically contingent upon the clinical acumen, experience, and interpretive prowess of healthcare practitioners. Given the intricate tapestry of clinical presentations that stroke weaves, the regrettable reality is that misdiagnoses and missed diagnoses often loom menacingly in the shadows of the diagnostic landscape. In recent years, however, there has emerged a burgeoning paradigm that promises to redefine the contours of stroke diagnosis and management—computer-aided diagnosis (CAD) technology.

This transformative technology represents a compelling frontier in the realm of medical imaging and diagnostic support. A litany of CAD technologies, imbued with machine learning and artificial intelligence capabilities, has burgeoned within the purview of medical practice. These technologies have soared in prominence, poised to play a pivotal role in enhancing diagnostic accuracy, bolstering the clinical acumen of healthcare practitioners, and, crucially, curtailing the incidence of misdiagnoses and overlooked cases.

To achieve an apex in the efficiency and precision of stroke diagnosis, the project at hand fervently espouses the integration of neural network technology within the expansive domain of machine learning. This innovative melding of computational and cognitive faculties seeks to illuminate the labyrinthine landscape of stroke classification, a task fraught with complexity and nuance. By harnessing the potent capabilities of neural networks, the medical community is poised to embark on a transformative journey—one characterized by heightened diagnostic acumen, enhanced patient outcomes, and a palpable diminishment in the inexorable burden that stroke imposes upon individuals and societies alike. It is, indeed, a noble endeavor, one that endeavors to combat the scourge of stroke with the full arsenal of modern technology and human ingenuity.

3.2 Relevant Technologies for Diagnosing Stroke

In recent years, the landscape of medical research has been dramatically transformed by the proliferation of machine learning, particularly within the domain of computer vision. Among the myriad of medical conditions that have come under the machine learning microscope, stroke has taken center stage due to its devastating impact on both mortality and disability rates. Researchers, healthcare practitioners, and data scientists have embarked on a collective journey to leverage the power of machine learning to enhance the diagnosis, treatment, and prevention of stroke. The overarching goal is to increase the likelihood of successful patient outcomes by providing expert-assisted diagnostic information and support.

One notable avenue of exploration in stroke research has been the application of traditional machine learning algorithms, with a particular emphasis on the utilization of support vector machines (SVMs). SVMs have a well-established reputation for their prowess in binary classification tasks, and as such, they have found a valuable niche in the realm of stroke management. A noteworthy example of this is the work by Bacchi and colleagues (2020), where SVMs were employed to assist in various aspects of stroke care.

The classical SVM model is undeniably adept at handling binary classification challenges, showcasing commendable performance even when dealing with relatively small datasets. However, the medical field, where the stakes are often matters of life and death, calls for solutions that are not only accurate but also scalable to accommodate the growing volume of medical data. While SVMs hold their own in scenarios characterized by modest data volumes, they face limitations when tasked with processing extensive datasets, a trend that is becoming increasingly prevalent with the rapid advancement of medical technology.

One of the key limitations inherent to the original SVM algorithm is its inherent inability to readily address multi-class classification problems. In the context of stroke, which encompasses a diverse spectrum of subtypes, such as ischemic, thrombotic, embolic, and transient ischemic attack, the ability to distinguish and classify these subtypes accurately is paramount. Physicians heavily rely on this nuanced classification for both diagnosis and treatment planning. SVM's binary nature poses a challenge when confronted with the intricate web of stroke subcategories, often necessitating cumbersome workarounds or complex one-vs-all strategies.

In response to these challenges, an alternative approach has gained prominence—the use of convolutional neural networks (CNNs). CNNs, celebrated for their remarkable performance in image analysis tasks, have found their way into the realm of medical imaging, holding immense promise for stroke diagnosis. However, the application of CNNs in stroke research is not without its complexities. Notably, there is a divergence in raw data types across different studies. While some researchers employ clinical CT images of the brain as the primary raw data source, others opt for alternative imaging modalities or data sources. Moreover, many of these studies grapple with relatively modest sample sizes, with training datasets comprising as few as 204 or 420 images.

The quest for more accurate, scalable, and multi-class classification models in the context of stroke diagnosis continues to evolve as a dynamic and high-impact field of research. Researchers are confronted with the complex interplay of medical data, technology, and the imperative to harness machine learning to improve healthcare outcomes in the challenging domain of stroke management. As technology advances and datasets continue to burgeon, the landscape remains ripe with opportunities to further refine and expand the application of machine learning in the battle against stroke, its multifaceted subtypes, and its profound consequences on the health and well-being of individuals and communities worldwide. This journey is marked by both challenges and breakthroughs, offering a beacon of hope for a future where stroke is more effectively diagnosed, treated, and prevented.

The quest for more accurate, scalable, and multi-class classification models in the context of stroke diagnosis continues to evolve as a dynamic and high-impact field of research. Researchers are confronted with the complex interplay of medical data, technology, and the imperative to harness machine learning to improve healthcare outcomes in the challenging domain of stroke management. As technology advances and datasets continue to burgeon, the landscape remains ripe with opportunities to further refine and expand the application of machine learning in the battle against stroke, its multifaceted subtypes, and its profound consequences on the health and well-being of individuals and communities worldwide. This journey is marked by both challenges and breakthroughs, offering a beacon of hope for a future where stroke is more effectively diagnosed, treated, and prevented.

3.3 Comparison of Neural Network Algorithms in Stroke Diagnosis

3.3.1 Data Acquisition and Preprocessing

In the first part, many standard neural network models are used to analyze and learn CT images, and the final accuracy is summarized and compared. Data was obtained from the stroke CT image dataset on the open data platform Kaggle (https://www.kaggle.com/datasets/afridirahman/brain-stroke-ct-image-dataset), which contained a total of 2501 brain CT images, among which 1551 were non-stroke brain CT images, and the other 950 were CT images of the stroke's brain. Of these, every 30 to 40 images were grouped, showing the results of a cross-sectional scan of different locations in a patient's brain. The learning and modeling process is carried out in Python 3.9 environment, and the editor is VS Code. The data image is shown in Fig. 3.1.

3.3.1.1 Normalization

The prediction model with strong generalization ability can be obtained from high-quality data. However, the original data collected in various ways is unsuitable for direct learning, so it is necessary to preprocess the data according to the specific task. Normalization is a common data preprocessing method that applies to the data of various models. By scaling or converting the original data, different characteristic values are adjusted to the close range (Krizhevsky et al., 2017). The advantage of normalization is that it can reduce the need for manual intervention and accelerate the convergence of the model. In this study, we used the relatively common Min–Max normalization. The method scales the value of the data between [0, 1] and [−1, 1], and the specific operation is as follows for each x_{ik} eigenvalue in each dimension x_i:

$$\hat{x}_{ik} = \frac{x_{ik} - \min(x_i)}{\max(x_i) - \min(x_i)}$$

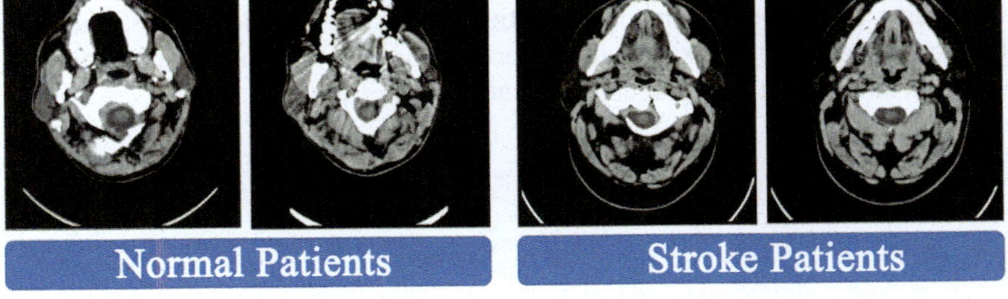

Fig. 3.1 Comparison of brain CT images of normal patients and stroke patients

Max (x_i) and min (x_i) are the maximum and minimum values of x_i, respectively.

For an image, the minimum value for each pixel point is 0, and the maximum value is 255. Size normalization for the image is divided by 255.0 for each pixel point.

3.3.1.2 Partitioning of Datasets

According to the ratio of label 7:3, partition the training set and the test set randomly. After division, the training set contains 1750 CT images, and the test set includes 751 CT images. The training set had 1092 normal CT images and 658 stroke images. The test set has 459 normal CT images and 292 stroke images. To ensure the repeatability of the training, the random seed is set to 0.

3.3.2 Modeling Using Multiple Convolutional Neural Networks

AlexNet

Convolutional neural network solves the problem of dataset size and model complexity faced by traditional machine learning algorithms. As one of the classic CNNs models, AlexNet, the champion of the ImageNet LSVRC-2010 competition, is a breakthrough. Its main framework consists of five convolution layers and three fully connected layers (as shown in Fig. 3.2). However, as an early convolutional neural network model, AlexNet's convolution kernel size (11 $\times$ 11) and step size (4 lattices) are relatively large, and no padding operation is performed before the convolution operation. As a result, the effect of feature extraction is not excellent compared with the new model in recent years.

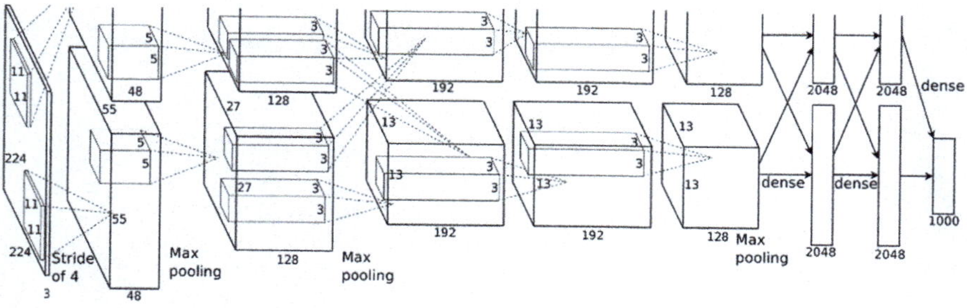

Fig. 3.2 Overall structure of AlexNet

VGG-16

VGG is a deeper convolutional neural network based on LeNet-5 and AlexNet. However, unlike AlexNet, VGG uses a smaller convolution kernel (3 × 3) and step size (1 lattice) and is more detailed for feature extraction. Although, in theory, multiple 3 × 3 convolution kernels have the same receptive field as a single large-sized convolution kernel, multiple small convolution kernels have more nonlinearity and fewer training parameters. Secondly, VGG tries to expand the depth of the network and proposes VGG-16 and VGG-19 (representing the depth of 16 and 19 layers, respectively). The deeper network represents the better learning ability and classification ability, and both VGG-16 and VGG-19 perform better than AlexNet's 8-layer deep network. In addition, VGG increases the number of convolution kernels after each maximal pooling to make up for the lost features of pooling. The specific structure is shown in Fig. 3.3.

The white part is a convolution layer, the red part is a pooling layer (maximum pooling is used), and the blue part is a fully connected layer, among which the activation functions of both the convolution layer and the fully connected layer use ReLu. The VGG16 has 16 layers, 13 convolution layers, and three thoroughly connected layers. After two convolutions with 64 convolution kernels for the first time, pooling is used once. After two convolutions with 128 convolution kernels for the second time, pooling is used again. After two times of convolution with three 512 convolution kernels, pooling is performed again, and finally, the full connection is performed three times. Figure 3.4 is an explanation of the model structure of VGG16.

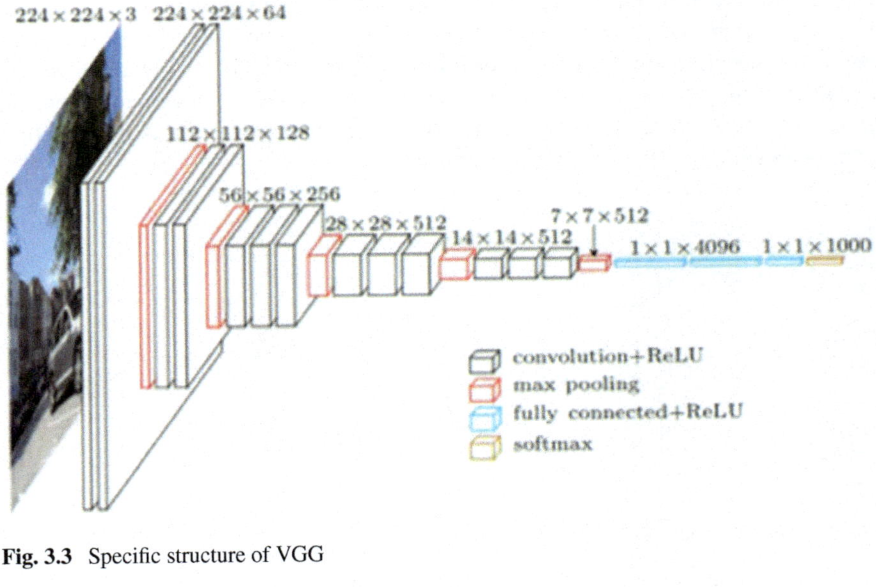

Fig. 3.3 Specific structure of VGG

ConvNet Configuration					
A	A-LRN	B	C	D	E
11 weight layers	11 weight layers	13 weight layers	16 weight layers	16 weight layers	19 weight layers
input (224 × 224 RGB image)					
conv3-64	conv3-64 **LRN**	conv3-64 **conv3-64**	conv3-64 conv3-64	conv3-64 conv3-64	conv3-64 conv3-64
maxpool					
conv3-128	conv3-128	conv3-128 **conv3-128**	conv3-128 conv3-128	conv3-128 conv3-128	conv3-128 conv3-128
maxpool					
conv3-256 conv3-256	conv3-256 conv3-256	conv3-256 conv3-256	conv3-256 conv3-256 **conv1-256**	conv3-256 conv3-256 **conv3-256**	conv3-256 conv3-256 conv3-256 **conv3-256**
maxpool					
conv3-512 conv3-512	conv3-512 conv3-512	conv3-512 conv3-512	conv3-512 conv3-512 **conv1-512**	conv3-512 conv3-512 **conv3-512**	conv3-512 conv3-512 conv3-512 **conv3-512**
maxpool					
conv3-512 conv3-512	conv3-512 conv3-512	conv3-512 conv3-512	conv3-512 conv3-512 **conv1-512**	conv3-512 conv3-512 **conv3-512**	conv3-512 conv3-512 conv3-512 **conv3-512**
maxpool					
FC-4096					
FC-4096					
FC-1000					
soft-max					

Fig. 3.4 Model structure of VGG16

Residual Network (ResNet)

A deeper and deeper network is constructed when it is discovered that a deeper network can improve the model's accuracy. Still, when the network becomes deeper, the accuracy sometimes does not increase but decreases. It has proved that building a deeper network is not a foolproof way to improve the model performance. The reason is the gradient disappearance or the result of gradient explosion and degradation (layer saturation, making the optimization difficult and the enormous error value). Based on these two problems, two strategies are proposed for the residual network. One is to use standard initialization in data preprocessing to solve the gradient disappearance or explosion problem. The other one introduces the residual structure (as shown in Fig. 3.5). The residual structure solves the degradation problem by introducing identity mapping and allowing some layers to be skipped. That is, for a layer with a poor learning effect, the model can skip the layer and go directly to the next layer

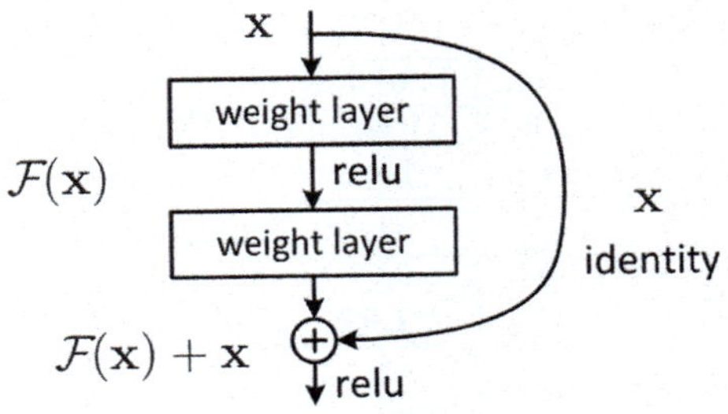

Fig. 3.5 Schematic diagram of residual structure

to prevent the result of the accuracy rate decrease so that increasing the depth can stably improve the accuracy rate.

GoogLeNet

As the first classical model using parallel network structure, GoogLeNet has provided many creative structures for building neural networks. The original goal was to solve the problem of many overfitting parameters that were over-fitted after the network was deepened. An Exception structure was proposed (several improved versions, such as v1-v4 and Inception-ResNet, and Inception-v1 was used in this study). The network has a fixed single running direction in the previous tandem networks (such as VGG and AlexNet). The size of each convolution kernel is fixed, so the feature extraction is also a single class, which makes the model's generalization ability poor. Inception-v1 uses convolution kernels of multiple sizes and concatenates their extracted feature maps together to improve the ability of multi-scale feature extraction (Fig. 3.6).

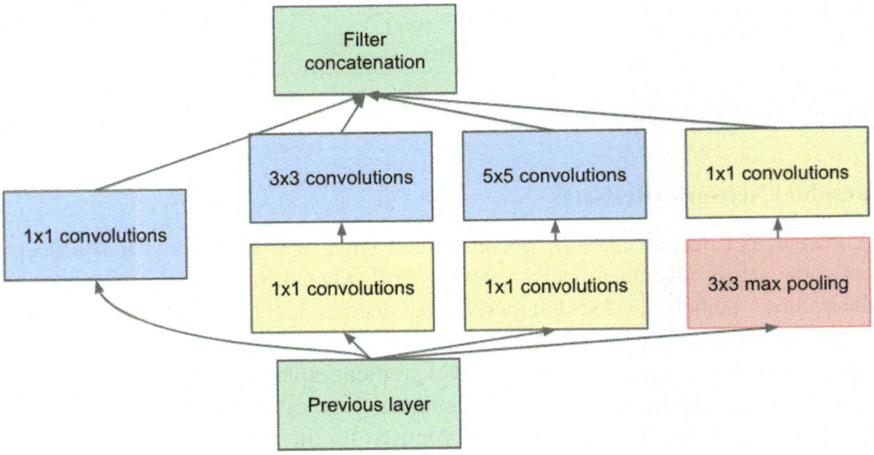

Fig. 3.6 Structure of exception-v1

Layers	Output Size	DenseNet-121	DenseNet-169	DenseNet-201	DenseNet-264
Convolution	112×112	7×7 conv, stride 2			
Pooling	56×56	3×3 max pool, stride 2			
Dense Block (1)	56×56	$\begin{bmatrix} 1 \times 1 \text{ conv} \\ 3 \times 3 \text{ conv} \end{bmatrix} \times 6$	$\begin{bmatrix} 1 \times 1 \text{ conv} \\ 3 \times 3 \text{ conv} \end{bmatrix} \times 6$	$\begin{bmatrix} 1 \times 1 \text{ conv} \\ 3 \times 3 \text{ conv} \end{bmatrix} \times 6$	$\begin{bmatrix} 1 \times 1 \text{ conv} \\ 3 \times 3 \text{ conv} \end{bmatrix} \times 6$
Transition Layer (1)	56×56	1×1 conv			
	28×28	2×2 average pool, stride 2			
Dense Block (2)	28×28	$\begin{bmatrix} 1 \times 1 \text{ conv} \\ 3 \times 3 \text{ conv} \end{bmatrix} \times 12$	$\begin{bmatrix} 1 \times 1 \text{ conv} \\ 3 \times 3 \text{ conv} \end{bmatrix} \times 12$	$\begin{bmatrix} 1 \times 1 \text{ conv} \\ 3 \times 3 \text{ conv} \end{bmatrix} \times 12$	$\begin{bmatrix} 1 \times 1 \text{ conv} \\ 3 \times 3 \text{ conv} \end{bmatrix} \times 12$
Transition Layer (2)	28×28	1×1 conv			
	14×14	2×2 average pool, stride 2			
Dense Block (3)	14×14	$\begin{bmatrix} 1 \times 1 \text{ conv} \\ 3 \times 3 \text{ conv} \end{bmatrix} \times 24$	$\begin{bmatrix} 1 \times 1 \text{ conv} \\ 3 \times 3 \text{ conv} \end{bmatrix} \times 32$	$\begin{bmatrix} 1 \times 1 \text{ conv} \\ 3 \times 3 \text{ conv} \end{bmatrix} \times 48$	$\begin{bmatrix} 1 \times 1 \text{ conv} \\ 3 \times 3 \text{ conv} \end{bmatrix} \times 64$
Transition Layer (3)	14×14	1×1 conv			
	7×7	2×2 average pool, stride 2			
Dense Block (4)	7×7	$\begin{bmatrix} 1 \times 1 \text{ conv} \\ 3 \times 3 \text{ conv} \end{bmatrix} \times 16$	$\begin{bmatrix} 1 \times 1 \text{ conv} \\ 3 \times 3 \text{ conv} \end{bmatrix} \times 32$	$\begin{bmatrix} 1 \times 1 \text{ conv} \\ 3 \times 3 \text{ conv} \end{bmatrix} \times 32$	$\begin{bmatrix} 1 \times 1 \text{ conv} \\ 3 \times 3 \text{ conv} \end{bmatrix} \times 48$
Classification Layer	1×1	7×7 global average pool			
		1000D fully-connected, softmax			

Fig. 3.7 DenseNet structure used for ImageNet

DenseNet

Based on ResNet, DenseNet believes that feature reuse may be a better method than identity mapping to prevent model degradation. Let each layer's inputs come from the outputs of all previous layers. This method allows features and gradients to be transferred more efficiently between layers. It mitigates the problem of gradient disappearance because each layer can capture the loss value of the previous layer (Fig. 3.7).

SqueezeNet

While various neural networks are pursuing higher accuracy, the authors of SqueezeNet want to make networks smaller (train fewer parameters to achieve the same accuracy) while achieving the same performance. Smaller networks have many advantages: First, for networks with distributed training, having a smaller number of networks can generally learn faster because the interaction required between servers decreases as parameters decrease; for autonomous driving, companies such as Tesla regularly copy new network models to customers' cars, and smaller networks can do this more frequently; and finally, for devices with smaller memory, smaller networks are easier to deploy to enlarge the application range of the network.

There are three strategies that SqueezeNet chose to scale down the network: replacing the smaller convolution kernel, using squeeze layers to reduce the input channels to the 3×3 convolution kernel, and downsampling later in the network so the convolution layer has a large activation map. SqueezeNet mainly consists of a Fire module, which includes a Squeeze (consisting of a bunch of 1×1 convolution kernels) part and an Expand (consisting of a set of 1×1 convolution kernels plus a group of 3×3 convolution kernels) part. It should be noted that, according to the second strategy, the number of channels in the Squeeze part of a Fire module should be less than the number of channels in the Expand part so that the number of channels

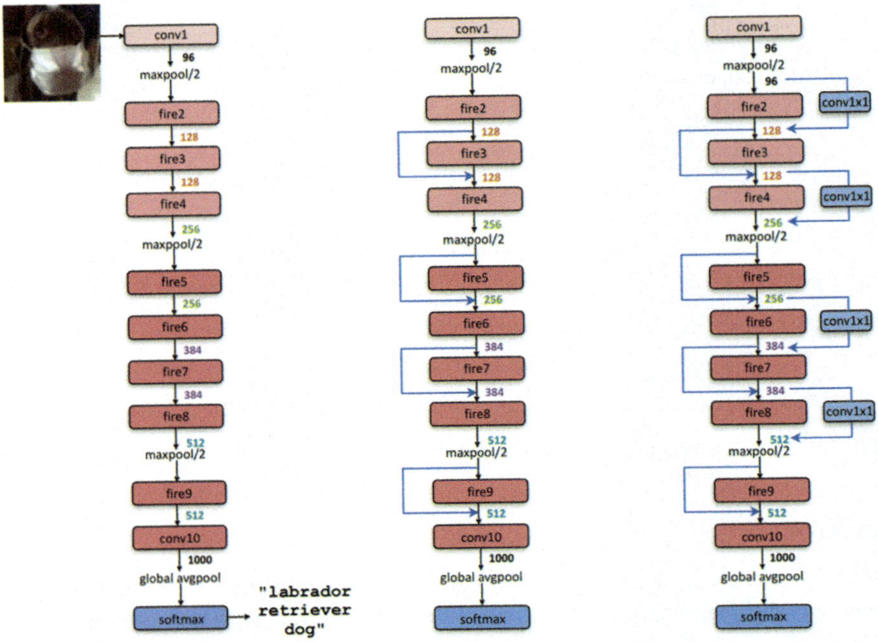

Fig. 3.8 Network architecture of SqueezeNet

input to the next 3 × 3 convolution kernel is reduced after the image passes through the squeeze layer. The network architecture of SqueezeNet is shown in Fig. 3.8.

3.3.3 *Comparison of Modeling Results*

3.3.3.1 Evaluation and Comparison Criteria

This study calculated accuracy, recall, precision, F1-score, the area under the ROC curve (AUC), and training speed was calculated as model evaluation and comparison criteria. The predicted results are true positive TP, false positive FP, false negative FN, and true negative TN. The true positive means that the patient is suffering from a stroke, and the predicted result is also a stroke; the true negative case means that the patient is normal, and the predicted result is also normal; the false positive case means that the true patient is normal, but the predicted outcome is stroke; the false negative refers to the patient has a stroke, but the predicted result is normal.

The accuracy rate represents the percentage of the number of images predicted correctly to the total number of images; the precision rate represents the rate at which the pictures predicted to be stroke were correctly predicted; the recall rate represents the proportion that was predicted to be correct in the images that were stroke; the

F1-score is calculated by the precision rate and the recall rate, which is equivalent to a comprehensive evaluation of the two results. Their calculation formulas are as follows, respectively:

$$\text{Accuracy} = \frac{\text{TP} + \text{TN}}{\text{TP} + \text{FP} + \text{FN} + \text{TN}}$$

$$\text{Precision} = \frac{\text{TP}}{\text{TP} + \text{FP}}$$

$$\text{Recall} = \frac{\text{TP}}{\text{TP} + \text{FN}}$$

$$\text{F1}_{\text{score}} = 2 \times \frac{\text{Precision} \times \text{Recall}}{\text{Presision} + \text{Recall}}$$

Calculation of each model standard is done by code: (taking the AlexNet model as an example)

```
import numpy as np
from klearn.metrics import precision_score, recall_score, f1_score, accuracy_
score
predict_x=model_alexnet.predict(test_X)
predict=np.argmax(predict_x,axis=1)
precision=precision_score(test_y,predict,average='binary')
recall=recall_score(test_y,predict,average='binary')
f1score=f1_score(test_y,predict,average = 'binary')
accuracy=accuracy_score(test_y,predict).
```

3.3.3.2 Training Parameter Details

Except for the network structure, the same parameters are used for each model. Firstly, to simplify the complexity of the model and facilitate the uniform input, all CT images were reduced from the original 650 × 650 to 224 × 224 pixels. Then the normalization operation was performed to improve the model training speed further. The optimizer adopts an Adam optimizer, a variation of gradient descent, which makes the learning rate of each iteration parameter within a specific range and the parameter adjustment relatively stable. The loss function uses the cross entropy (sparse_categorical_crossentropy) for multiple classifications. The batch size is 32, each model is trained for 50 rounds, the proportion of the verification set is 0.2, and the learning is finished before the loss value of the verification set has no apparent change.

3.3.3.3 Model Evaluation and Summary

The learning results are as in Table 3.1. All five classical models on each index have completed the learning and achieved high accuracy. However, VGG-16 failed to converge after additional parameter adjustment because of the model complexity, so the indexes were not recorded. Although the accuracy of each model is almost the same, the learning speed is quite different. As the earliest neural network model, AlexNet achieves the fastest speed in the case of completing the learning by a simple structure, while SqueezeNet, optimized for deployment and depth, also has a quicker speed (Fig. 3.9).

Table 3.1 Performance indicators of each model on the test set

	AUC	Accuracy rate	Precision rate	Recall rate	F1 score	Training speed
AlexNet	0.997	0.981	0.976	0.976	0.976	6 s/epoch
ResNet	0.996	0.980	0.986	0.962	0.974	25 s/epoch
GoogLeNet	0.996	0.969	0.969	0.952	0.960	17 s/epoch
DenseNet	0.998	0.979	0.960	0.986	0.930	36 s/epoch
SqueezeNet	0.997	0.980	0.983	0.966	0.974	8 s/epoch

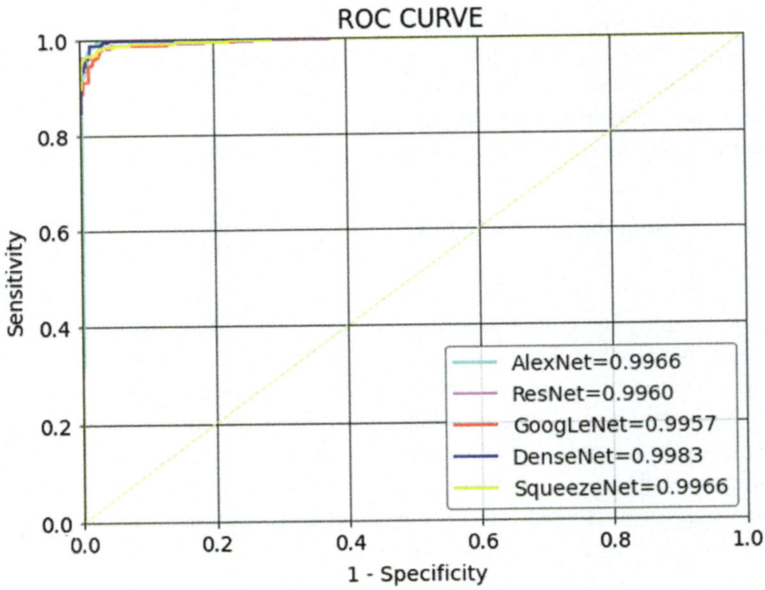

Fig. 3.9 Prediction result ROC curve

3.3.4 *Implementation of Stroke Diagnosis System*

Based on the experimental results, we chose AlexNet as the final network model for the assistant analysis of brain CT in this stroke diagnosis system because AlexNet is lightweight and still maintains high accuracy. Finally, code is written to read the picture address and output the result of the model prediction.

```
def obtain_image(path):
input_image=path
image=cv2.imread(input_image)
image=cv2.cvtColor(image,cv2.COLOR_BGR2RGB)
image=cv2.resize(image,(224,224))
image=np.array([image])
return image/255.0
def predict_class(image):
predict=model_alexnet.predict(image)
pre_class=np.argmax(predict,axis=1)
return 'stroke'ifpre_class==1else'normal'
image_path=input('enter picture address:')
image=obtain_image(image_path)
predict_class(image)
```

3.4 Conclusion and Limitation

In recent years, the realm of medical research has been irrevocably altered by the proliferation of machine learning, particularly within the subfield of computer vision. This transformative wave has swept across various medical conditions, but perhaps nowhere has its impact been more profound than in the context of stroke, a medical emergency of monumental significance. Stroke, often colloquially referred to as apoplexy, represents a cluster of cerebral disorders characterized by the catastrophic rupture or insidious blockage of cerebral blood vessels, leading to a disruption in the delicate dance of blood flow within the brain. It is an affliction that predominantly haunts the middle-aged and elderly populace, with its incidence escalating in lockstep with the inexorable march of time. As a result, it has laid claim to the dubious distinction of being the second leading cause of mortality on a global scale and stands as a ubiquitous specter in the annals of disability statistics.

The magnitude of stroke's impact becomes starkly evident when we examine data published by the World Stroke Organization (WSO) in its 2022 Global Stroke Fact Sheet. It is revealed that a staggering 80% of the 12 million new stroke cases registered worldwide are attributed to individuals aged 49 and above. This dire statistic underscores the inescapable reality that stroke's menacing shadow looms ever larger with the progression of age. Furthermore, a sobering portrait of global stroke epidemiology emerges, as one in every four individuals aged 25 or older is destined to confront the ominous specter of stroke during their lifetime.

The impact of stroke is not confined to global statistics; it strikes with particular acuity in China, a burgeoning superpower with the world's largest population. The China National Stroke Center Report of 2020 paints a disquieting picture, revealing that in the same year, an astonishing 17.8 million individuals in China aged 40 and above grappled with the relentless clutches of stroke. This devastating tidal wave included 3.4 million new cases that cruelly materialized, alongside a heart-wrenching tally of 2.3 million stroke-related fatalities. This unassailable testament to the overwhelming scale of stroke's impact underscores the disquieting ubiquity of this affliction, which touches the lives of countless individuals across the nation.

Diagnosing stroke presents a formidable clinical challenge, one that hinges entirely on the clinical experience and medical image interpretive prowess of healthcare practitioners. Given the intricate tapestry of clinical presentations that stroke weaves, it is an unfortunate reality that misdiagnoses and missed diagnoses often loom menacingly in the shadows of the diagnostic landscape. It is here that computer-aided diagnosis (CAD) technology emerges as a beacon of hope.

CAD technology, a burgeoning paradigm in the realm of medical imaging, has rapidly ascended as a research hot spot, offering the promise of improved diagnostic accuracy and reduced instances of missed diagnoses. By harnessing the power of machine learning, CAD systems aim to enhance the efficiency and accuracy of CT image diagnosis, particularly in the context of stroke. These systems are designed to provide valuable diagnostic insights and support to healthcare professionals, thereby increasing the likelihood of successful patient outcomes.

The application of CAD technology in stroke diagnosis has ushered in a new era of medical practice. Researchers and data scientists have explored various machine learning algorithms, with particular attention to the venerable support vector machines (SVMs). SVMs, renowned for their prowess in binary classification tasks, have been harnessed as valuable tools in the realms of stroke prevention, diagnosis, treatment, and pre-screening, as demonstrated by the work of Bacchi et al. (2020).

However, even as SVMs showcase commendable performance in scenarios characterized by modest data volumes, they grapple with scalability challenges when confronted with vast datasets—a phenomenon increasingly prevalent in the wake of rapid technological advancement. SVM's inherent limitation in handling multi-class classification problems further complicates their application in stroke diagnosis, which often requires nuanced classification among various stroke subtypes (e.g., ischemic, thrombotic, embolic, and transient ischemic attack).

In response to these challenges, researchers have begun to explore the potential of convolutional neural networks (CNNs) as an alternative approach to disease classification, particularly in medical imaging. CNNs, celebrated for their prowess in image analysis tasks, hold immense promise for stroke diagnosis. However, one notable caveat lies in the variability of raw data types across different studies. While some researchers employ clinical CT images of the brain as their primary data source, others opt for alternative imaging modalities or data sources. Moreover, many of these studies contend with relatively modest sample sizes, with training datasets comprising as few as 204 or 420 images.

The pursuit of more accurate, scalable, and multi-class classification models in the context of stroke diagnosis continues to evolve as a dynamic field of research. Researchers grapple with the complex interplay of medical data, technology, and the imperative to harness machine learning to improve healthcare outcomes in the challenging domain of stroke management. As technology advances and datasets continue to burgeon, the landscape remains ripe with opportunities to further refine and expand the application of machine learning in the battle against stroke, its multi-faceted subtypes, and its profound consequences on the health and well-being of individuals and communities worldwide. This journey is marked by both challenges and breakthroughs, offering a beacon of hope for a future where stroke is more effectively diagnosed, treated, and prevented.

A comprehensive comparison of the results obtained in the stroke diagnosis system reveals some intriguing insights into the application of machine learning models for diagnosing strokes based on brain CT images. Notably, it becomes apparent that achieving accurate stroke diagnosis does not necessarily hinge on deploying excessively complex models or pursuing meticulous optimization. Even a decade-old architecture like AlexNet has demonstrated commendable performance in this context. However, contemporary models often bring more advantages than drawbacks when seeking higher accuracy through increased complexity.

Several crucial factors come into play when considering the complexity of machine learning models for stroke diagnosis. First and foremost is the computational cost associated with training overly complex models. This cost manifests in terms of time and computational resources, and it is worth noting that different models exhibit varying training times per epoch under identical environments. The swiftness of convergence directly impacts the number of epochs required to attain the desired accuracy level.

Furthermore, the use of overly complex models can give rise to issues such as overfitting and convergence difficulties, particularly in scenarios characterized by relatively simple classification problems. Overfitting occurs when a model becomes excessively tailored to the training data, resulting in reduced generalization to new, unseen data. Balancing model complexity and generalization capability is paramount, especially in medical contexts where robustness and reliability are essential.

Additionally, the future landscape of machine learning deployment may necessitate smaller, more efficient models suitable for deployment on constrained devices and chips. This includes scenarios where machine learning models are integrated into human-implanted chips, requiring compact model sizes and ease of deployment. The emergence of architectures like SqueezeNet in recent years underscores the importance of accommodating such deployment scenarios.

It is essential to highlight that various neural network models were trained under uniform conditions and parameters in this project. This uniformity of approach limits the potential performance of some networks, such as VGG and DenseNet, which are renowned for their ability to enhance model performance through deep networks. These architectures have excelled in multi-classification problems, such as the renowned ImageNet competition.

The overarching aim of this project is to delve into the application of diverse neural network models in the realm of stroke diagnosis, thereby contributing to the body of knowledge and validation in this critical domain. By utilizing a broader spectrum of open data and an array of neural network architectures, the project seeks to unravel the intricacies of the problem and determine the varying performance demands of different models.

When it comes to classifying stroke patients, maintaining an accuracy threshold of approximately 98%, it is evident that a shallower neural network model can be embraced to enhance equipment portability and deployment speed. Conversely, in specialized medical diagnostic contexts where larger datasets and heightened accuracy are paramount, deeper neural network models like residual networks (ResNets) and GoogLeNet can be harnessed to explore enhanced accuracy and multi-classification capabilities.

In summary, this study harnesses the power of the gray-level co-occurrence matrix (GLCM) in conjunction with BP neural network technology to provide auxiliary diagnosis for brain CT images in the context of computer-aided diagnosis for stroke. The benefits are manifold, including lightening the workload of medical professionals, furnishing critical auxiliary diagnosis information, and effectively mitigating the occurrence of misdiagnoses and missed diagnoses. This application holds significant promise and inherent value, despite the limitations imposed by the unavailability of clinical data from hospitals and the potential mismatch between publicly available datasets and real-world clinical variability. It is a step forward in the journey to enhance stroke diagnosis and improve patient outcomes through the integration of machine learning and medical imaging.

References

Bacchi, S., Zerner, T., Oakden-Rayner, L., Kleinig, T., Patel, S., & Jannes, J. (2020). Deep learning in the prediction of ischaemic stroke thrombolysis functional outcomes. *Academic Radiology, 27*(2), e19–e23. https://doi.org/10.1016/j.acra.2019.03.015

Corrigendum to: World Stroke Organization (WSO): Global Stroke Fact Sheet 2022. (2022). *International Journal of Stroke: Official Journal of the International Stroke Society, 17*(4), 478. https://doi.org/10.1177/17474930221080343

Greenberg, D. A., Aminoff, M. J., & Simon, R. P. (2002). *Clinical neurology*. McGraw-Hill.

Krizhevsky, A., Sutskever, I., & Hinton, G. E. (2017). ImageNet classification with deep convolutional neural networks. *Communications of the ACM, 60*(6), 84–90. https://doi.org/10.1145/3065386

Ye Xia Graduated from University of Electronic Science and Technology of China with a master's degree in software engineering. He received his master's degree in computer engineering from California State University in 2018. In 2019, he joined Beijing Royal Palace School as computer science and AP Computer Principal Teacher.

Chapter 4
Application of Face Detection for Learning Engagement in Classroom

Jiaxuan Chen, Jiaming Zhao, Yuxuan Deng, Lan Wu, and Ye Gao

Abstract Integrating artificial intelligence (AI) and education has led to the widespread application of facial recognition technology in various fields, including classroom teaching. In this project, AI methods are used to acquire facial expression data of students to create an effective classroom state evaluation mechanism that assists teachers in improving the quality and efficiency of classroom teaching. The research involves analyzing the significance of the application of face detection technology in classroom teaching, constructing a face detection dataset based on the classroom environment, and identifying critical methods used in classroom teaching application systems. The proposed face detection system for teaching in the classroom improves traditional face detection algorithms by optimizing the network structure and implementing a lightweight classroom monitoring program to evaluate each student's classroom status. The real-time detection of students' facial expressions provides teachers with timely feedback to enhance teaching quality and learning engagement in the classroom.

Keywords Dynamic face detection · Real-time control · Learning engagement

---

The Second Branch School of the High School Affiliated to Renmin University of China is a public school affiliated to the Haidian District Education Commission and a member school of the Joint General School of the High School Affiliated to Renmin University of China. It is a practitioner of the balanced quality education policy vigorously promoted by the Haidian District Education Commission. In 2020, the school won the first batch of new brand schools in Haidian District.

J. Chen (✉) · J. Zhao · Y. Deng · L. Wu · Y. Gao
The Second Branch School of the High School Affiliated to Renmin, University of China, Beijing 100097, China
e-mail: chenjiaxuan2006@163.com

R. Huang et al. (eds.), *A Collection of AI Innovations by Chinese Teenagers*, Smart Computing and Intelligence, https://doi.org/10.1007/978-981-97-3316-3_4

Highlight

This project explores the integration of artificial intelligence and education to improve the quality and efficiency of classroom teaching. Our proposed face detection system provides real-time feedback to teachers on students' facial expressions, enhancing learning engagement and teaching quality.

How does the face detection system work to evaluate students' classroom status?

The proposed face detection system provides real-time detection of students' facial expressions, which can be used to evaluate their classroom status. The system is designed to detect multiple faces in a classroom environment and can accurately detect faces even under interference conditions such as facial change, illumination condition, and face shift. The system uses critical technology and algorithm theory of classroom state monitoring systems based on expression analysis. The real-time detection of students' facial expressions provides teachers with timely feedback to enhance teaching quality and learning engagement in the classroom.

What methods were used to construct the face detection dataset based on the classroom environment?

To construct the face detection dataset based on the classroom environment, the project team collected data in a virtual classroom environment and used human face information to create a multi-face dataset under the classroom environment of face detection and face detection. The team reconstructed the 3D face model, obtained the face representation through the deep neural network, and then tried to obtain high accuracy by training on many datasets. However, the training data was limited during the implementation of this project. Therefore, the team built a batch of real classroom datasets and enhanced the data. The face datasets of different classes were constructed for the two processes of face detection and face detection and marked them, respectively. The existing public face datasets, such as FDDB and LFW, were not suitable for face detection research in a classroom environment, so the project needed to build a face dataset based on classroom teaching.

How does the integration of artificial intelligence and education improve classroom teaching?

The integration of artificial intelligence and education can improve classroom teaching in several ways. In this project, an artificial intelligence method is used to acquire the facial expression data of the students from the classroom. A deep-learning convolution neural network DCNN, an OpenCV computer vision database, and a Google MediaPipe Face Mesh are used to analyze and judge the classroom state of the students to form an effective classroom state evaluation mechanism to assist teachers in improving the quality and efficiency of classroom teaching. The proposed face detection system provides real-time feedback to teachers on students' facial expressions, enhancing learning engagement and teaching quality. By using artificial intelligence to monitor students' facial expressions, teachers can better understand their students' learning status and adjust their teaching methods accordingly. This can lead to more effective teaching time, improved overall quality of talent cultivation, and increased economic growth.

4.1 Project of Appling Face Detection Technology in Classroom

4.1.1 Background

The continuous advancement of artificial intelligence technology has ushered in a new era of innovation across various sectors, with education being no exception. Within the expansive field of artificial intelligence, face detection technology has emerged as a pivotal component of computer vision, finding application in a multitude of domains. Its versatility is evident in applications such as security systems with facial recognition capabilities, access control systems used for community management, and the implementation of face-based check-in and attendance systems in daily office routines. However, one of the most transformative arenas where face detection technology is making waves is within the realm of education.

The classroom, as the quintessential locus of learning and interaction between educators and students, has historically occupied a central role in the educational landscape. It is a space of immense importance, both for educational institutions and society at large. Classroom teaching remains the cornerstone of knowledge dissemination, where teachers impart essential lessons to students while fostering communication and engagement. However, traditional classroom teaching has always grappled with a fundamental challenge—the heterogeneous nature of student learning.

In a typical classroom, students exhibit varying degrees of comprehension and engagement with the subject matter. The teacher, tasked with guiding this diverse group, faces the formidable challenge of simultaneously monitoring and addressing the individual learning needs of each student in real time. Furthermore, accurately gauging whether every student has attained mastery of the lesson's key concepts remains a daunting task. Beyond verbal communication and group discussions, students' facial expressions can offer subtle yet valuable cues about their acceptance of the classroom knowledge. These non-verbal cues, often overlooked, provide a window into the students' emotional and cognitive states.

Classroom teaching, in essence, is a multifaceted process where teachers constantly seek points of convergence between their teaching strategies and the diverse needs of their students. In the context of a one-to-many classroom environment, where a single teacher instructs a group of students, the variability in individual learning progress and efficiency becomes increasingly pronounced. Some students may grasp concepts quickly, while others may lag behind. When students encounter questions or uncertainties, the opportunity to seek immediate clarification from the teacher is often constrained, potentially leading to a decline in teaching quality.

Recognizing these challenges, this project endeavors to harness the power of artificial intelligence to acquire facial expression data from students within the classroom. This data is then subjected to a sophisticated analysis using a deep-learning convolutional neural network (DCNN), the OpenCV computer vision database, and Google MediaPipe Face Mesh technology. The overarching objective is to develop an effective classroom state evaluation mechanism that empowers teachers to enhance the quality and efficiency of their teaching.

The application of artificial intelligence in this educational context holds immense promise. By monitoring students' facial expressions, the system can provide invaluable real-time feedback on their engagement, comprehension, and emotional state during lessons. This feedback is not a mere collection of data but a dynamic resource that can empower teachers to adapt their teaching methods on the fly. When a student's furrowed brow or puzzled expression is detected, it serves as a signal to the teacher that intervention or clarification may be necessary.

Moreover, the AI-based facial expression analysis introduces an element of personalization into the teaching process. It enables teachers to identify students who may be struggling or disengaged, allowing for timely support or clarification to be provided. Additionally, the technology facilitates the identification of patterns and trends in student responses. This data-driven insight equips educators with the means to make informed decisions about their teaching strategies, tailoring them to the unique needs of their students.

In summary, the integration of artificial intelligence and facial expression analysis into the classroom environment represents a profound evolution in the field of education. It offers teachers a set of tools that can potentially revolutionize their teaching practices. By providing real-time insights into student engagement and comprehension, it paves the way for more effective and tailored instruction. As technology continues to advance, its role in education is destined to become increasingly indispensable, creating engaging and impactful learning experiences for students and educators alike. The journey of innovation in education is ongoing, with the promise of continued enhancements and refinements in the quest to unlock the full potential of every student.

The problems to be solved in the project include:

(1) To study the background and significance of the applications of face detection technology in classroom teaching.
(2) To understand the research progress of face detection at home and abroad, analyze the shortcomings of the face dataset, such as LFW, and try to construct the dataset of face detection research based on the classroom teaching environment.
(3) To study and identify the critical methods used in classroom teaching application systems based on face detection systems.
(4) For the problem of classroom state classification, a classification model of the system is designed based on facial features.
(5) To design a classroom teaching application system based on the research of face detection, recognition, and feature classification. Real-time detection of students' facial expressions will provide feedback to teachers timely when the students are found to be in a bad classroom state or have doubts about the classroom content to improve the teaching quality and classroom efficiency.

It is used the critical technology and algorithm theory of classroom state monitoring systems based on expression analysis. The traditional face detection algorithm cannot meet the current application requirements because of its shortcomings, such as a large amount of computation and poor real-time performance, so a face detection system construction based on classroom teaching is proposed. The convolution operation in the traditional CNN model is improved, and the network structure of the model is optimized. Design and implement a lightweight classroom monitoring program to make each student's classroom status visible and evaluable.

4.1.2 *Project Impacts and Benefits*

The following section including the impacts and benefits of this project, which demonstrate the project's positive contributions in social, educational, and generalizable realms.

4.1.2.1 For Social Impacts

Education is the foundation of the nation and a strong country. The development of a country needs all kinds of talents. The talents need to be cultivated from an early age. Classroom education in primary and secondary schools is the most critical education method. Teachers in primary and middle schools can achieve the aim of cultivating talents by imparting knowledge to students in the classroom. Still, one-to-many on-site teaching is generally adopted in the existing classroom education mode. Due to the differences in the mastery and absorption of knowledge in the learning process and the different concentrations of each student in class, the teacher cannot grasp the state of each student in the real-time class and cannot judge whether the student has mastered or absorbed the knowledge. As a result, classroom efficiency is not high enough, and some students cannot achieve the best learning effect. Compared with other biological recognition technology, face detection technology is more non-invasive and can guarantee classroom teaching to the greatest extent. Therefore, we designed the project, which is based on dynamic face detection technology, and developed a set of intelligent teaching assistant systems to help teachers know the learning state of each student in real time to realize accurate teaching, improve classroom efficiency, and cultivate more outstanding talents for the country.

4.1.2.2 For Teaching Efficiency and Quality

In traditional classroom settings, teachers face the formidable challenge of simultaneously instructing and managing a diverse group of students. Each student's learning journey is unique, with varying degrees of comprehension and engagement with the subject matter. This heterogeneity necessitates individualized attention and support, which can be a resource-intensive endeavor for educators. However, the conventional classroom environment often lacks the tools and resources to effectively monitor and address these diverse learning needs.

The intelligent teaching assistant system developed in this project represents a transformative solution to this challenge. It harnesses the power of face detection technology to provide real-time insights into students' emotional and cognitive states during lessons. By capturing and analyzing students' facial expressions, the system can decipher their level of engagement, comprehension, and potential areas of confusion. This wealth of information serves as a dynamic feedback mechanism for teachers, empowering them to tailor their teaching methods in response to the students' needs.

The benefits of this intelligent teaching assistant system are manifold. First and foremost, it liberates teachers from the burden of constantly monitoring and managing the classroom, allowing them to focus more on the core task of teaching. This newfound efficiency not only saves time but also enhances the overall teaching experience. Teachers can allocate their attention more strategically, addressing the specific needs of students who require additional support or clarification.

Moreover, the system optimizes the utilization of precious teaching time. By providing timely feedback on student engagement and comprehension, it enables teachers to make real-time adjustments to their teaching methods. This adaptability ensures that teaching is always aligned with the students' current level of understanding, ultimately leading to more effective and efficient learning experiences.

The impact of the intelligent teaching assistant system extends beyond individual classrooms. It has the potential to be deployed widely across primary and secondary schools, transforming cutting-edge technology into practical educational tools. By bridging the gap between science and application, this system not only enhances educational quality but also generates substantial economic value.

The economic potential lies in the creation of a marketable product that can be adopted by schools worldwide. The system's ability to enhance classroom efficiency and teaching effectiveness can make it an indispensable tool for educators. Schools and educational institutions stand to benefit significantly from the improved educational outcomes and streamlined teaching processes facilitated by this technology. This, in turn, can lead to increased enrollment, improved academic performance, and enhanced reputation for educational institutions.

In summary, the intelligent teaching assistant system developed in this project is poised to redefine the educational landscape. By harnessing the capabilities of face detection technology, it empowers educators with valuable insights into their students' learning experiences. This innovation promises to elevate classroom efficiency, optimize teaching methods, and unlock the full potential of each student. Furthermore, its potential for widespread adoption positions it as a valuable product with the capacity to bring about substantial economic growth while advancing the quality of education. The future of education is brighter and more dynamic with the integration of intelligent teaching assistant systems, setting the stage for a new era of learning and growth.

4.1.2.3 For Learning Engagement

The intelligent teaching assistant system, born out of this project, stands as a testament to the boundless potential of artificial intelligence and face detection technology. Its transformative capabilities extend far beyond the confines of the classroom, offering a multifaceted solution that enhances education, streamlines administrative tasks, and promotes public safety. Let us delve deeper into the myriad applications and ramifications of this groundbreaking innovation:

(1) Campus Teaching and Management:

Classroom Discipline Maintenance: The system serves as an ever-watchful guardian of classroom discipline. By continuously monitoring students' actions, behaviors, and expressions, it enables teachers to promptly address disruptions or unusual activities. This proactive approach fosters a conducive and orderly learning environment, enhancing the overall teaching and learning experience.

Student Attendance Management: Tedious manual attendance-taking becomes a relic of the past with the system's automated attendance tracking. It meticulously records students' presence, providing an accurate and efficient alternative to traditional methods. This not only saves valuable teaching time but also minimizes the risk of errors in attendance records.

Personnel Access Control: Security on campus receives a significant boost as the system takes charge of personnel access control. By verifying the identities of individuals entering or exiting school premises, it fortifies security measures and ensures that only authorized personnel gain access to sensitive areas.

Canteen Management: Chaotic queues and time-consuming payment processes in school canteens are a thing of the past. The system streamlines meal settlement by enabling automatic payments, reducing waiting times, and simplifying the overall dining experience for students.

Library Access Control: The system extends its reach to library access control. It ensures that only authorized individuals can enter the library and borrow books, effectively managing library resources and enhancing security.

Book Borrowing Registration: Borrowing books from the library becomes a seamless and efficient process as the system automates the registration procedure. Students and staff benefit from expedited book borrowing, streamlining their access to educational resources.

(2) Online Learning Enhancement:

Revolutionizing Online Classrooms: Originally designed to support home-based online learning during the pandemic, the system possesses untapped potential in the realm of online education. By continuously monitoring students' actions, expressions, and engagement levels in real time, it empowers educators to gauge and respond to students' needs promptly. This adaptive approach revolutionizes online learning, fostering a more interactive, efficient, and productive virtual learning environment.

(3) Public Administration and Safety:

Personal Identification: The dynamic face detection technology becomes a stalwart ally in personal identification. Whether in virtual meetings or at critical access points, the system ensures that only authorized individuals gain entry, bolstering security and minimizing the risk of unauthorized access.

Baby Care: In the realm of infant care, the system serves as a watchful guardian. Nursing staff can monitor the status of infants in real time, providing timely attention and care to ensure their well-being and safety.

Public Safety: In bustling and crowded public spaces like customs checkpoints, airports, train stations, and shopping malls, the system stands as a vigilant sentinel. It identifies and flags abnormal individuals through face detection, serving as an invaluable tool in maintaining public safety and security. Potential threats or issues can be swiftly identified and addressed, ensuring the safety of all.

The intelligent teaching assistant system and the dynamic face detection technology represent an exquisite fusion of innovation and practicality. Their reach is vast, extending across the educational, administrative, and public safety domains. These technologies usher in convenience, efficiency, and safety guarantees for education, teaching, extracurricular activities, and various facets of people's work and daily lives.

As we gaze into the future, the potential for these intelligent systems appears limitless. They promise to connect us in unprecedented ways, enhance our security, and optimize our efficiency. With each passing day, the boundaries of what is possible continue to expand, and the stage is set for a world where technology serves as a steadfast ally, empowering us to achieve our fullest potential in education, work, and life.

4.2 Overall Project Design

To address the problems proposed in this project, the project team proposed to build an application system based on the dynamic face detection technology in the classroom and tried to enrich and improve the dataset, form expression model construction, conduct anomaly analysis on the collected data, do algorithm model construction by combining the big data platform, analyze the differences of various expressions, classify the characteristics of different facial expressions, summarize the algorithm of expression recognition, and put forward the potential application of expression recognition, challenges, and future research directions. The solutions to the core issues are as follows.

4.2.1 Design of Multi-Face Detection Scheme in Classroom Teaching

Currently, most face detection technologies are based on a single face. Still, for a classroom teaching environment, multi-face detection is needed, and the accuracy of face detection can be high under the interference conditions such as facial change, illumination condition, and face shift. The method adopted in this project is to reconstruct the 3D face model, obtain the face representation through the deep neural network, and then try to obtain high accuracy by training on many datasets. However, the training data is limited during the implementation of this project. Therefore, we try to build a batch of real classroom datasets and enhance the data. The face datasets of different classes are constructed for the two processes of face detection and face detection and marked them, respectively.

Because the existing public face datasets, such as Face Detection Data Set and Benchmark (FDDB), Labeled Faces in the Wild Home (LFW), and so on, although the data volume is enormous, they all come from faces in different environments. The number of each category is unequal, and the whole picture is mostly a single face, which is not suitable for face detection research in a classroom environment. Therefore, the project needs to build a face dataset based on classroom teaching. By collecting the data in the virtual classroom environment and using human face information, a multi-face dataset under the classroom environment of face detection and face detection is created to improve face detection accuracy.

The method comprises the following steps: First, collecting face picture data for all students, defining the composition of the picture data, including the storage form of the data, the number of faces contained in the picture, the angle, the time, and so on. Then, extracting the detailed features of the faces of the students through self-generating samples and network data and comparing the detailed features with the ones in the database to determine the identity of the students. In this project, we try to use two algorithms for the face detection experiment.

4.2.1.1 Establish Dataset for Face Detection

In the preliminary data, 20 facial pictures are collected firstly, 68 feature point arrays corresponding to the 20 pictures are collected, the file named face_features.txt (i is a number from 01 to 20) is saved in the same directory, and the mean feature array face_feature_mean.txt is calculated through the twenty features. Then, face features are extracted based on the dlib library and are captured in the video stream and saved the face features in a picture file with a size of 64 × 64. In image acquisition, please note: try to keep in a roughly fixed light range and manually remove the exposure and dark images. In addition, in this acquisition, to ensure the definition of the camera, the same equipment shall be used to collect the dataset for face detection (Fig. 4.1).

We use dlib and OpenCV programming for face detection because both CPP and Python interfaces are compatible with cooperation among our project team members.

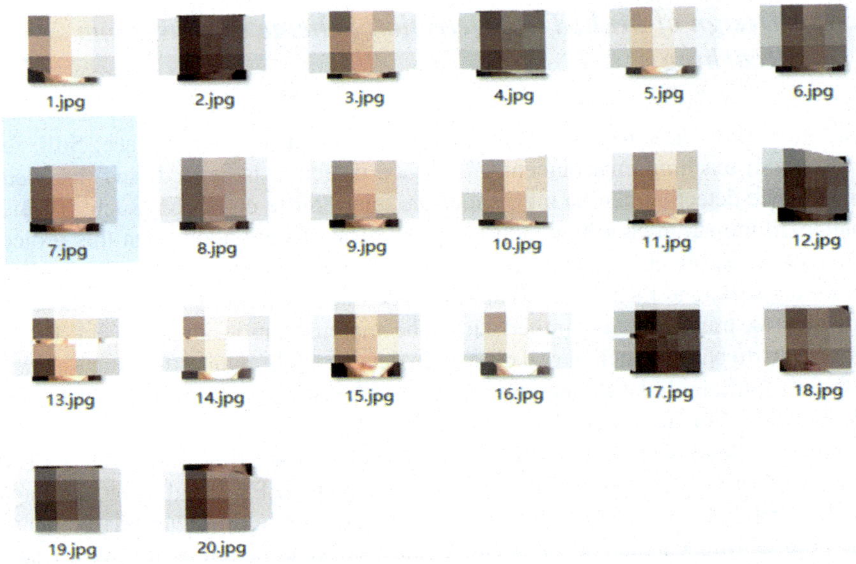

Fig. 4.1 Run results

However, it is found in video face detection that the real-time performance is not good and that there is some jamming phenomenon. In addition, because most of the faces in the photos in the classroom are angled and do not have front faces, the main principle of dlib and OpenCV face detection is to position the key points of the face to determine whether there is a face. This method must expose the key points of the detected faces to carry out location detection, so the non-positive face pictures cause great resistance to the detection. In the classroom teaching environment, most multi-face do not appear as full front faces. Therefore, there are some practical problems in multi-face detection with shielding and angle deviation. In the actual test, the test result is not ideal. The face can be detected under an excellent front angle and lighting, but the face in the side, shielding, and poor light environment cannot be detected. Therefore, we try another method for face detection.

4.2.1.2 Comparison and Selection of Implementation Method

Within the framework of this project, the development team meticulously considered the balance between algorithmic computational load and real-time performance. Their thoughtful approach led to the design and implementation of a sophisticated facial expression model centered around 478 + 10 key facial points. Here is an in-depth exploration of how this model operates and achieves its remarkable capabilities:

(1) Data Capture and Processing:

The foundation of the model lies in its ability to capture and process facial data in real time. This is achieved using the function, which captures data from the camera feed.

(2) Leveraging OpenCV for Facial Shape and Motion Perception:

OpenCV, a powerful computer vision library, is harnessed to perceive the shape and motion of the face. This forms the foundational understanding of facial expressions.

(3) MediaPipe Face Mesh:

The project leverages the capabilities of MediaPipe Face Mesh, a technology that extracts 478 facial coordinates from a single frame. This rich dataset becomes the cornerstone for real-time facial expression analysis.

(4) Real-time Hand-sensing Function:

Beyond facial expression analysis, MediaPipe Face Mesh also offers real-time hand-sensing capabilities. This additional functionality enhances the overall utility of the system.

(5) Expression Calculation Based on Key Point Coordinates:

The heart of the model lies in its ability to calculate facial expressions based on the coordinates of key facial points. These key points are derived from the MediaPipe Face Mesh technology.

(6) Evolution of Relative Key Point Positions:

The primary method employed is rooted in the analysis of 478 key point coordinates. It relies on the close positional relationships between these key points to calculate the shape of various facial expressions. By monitoring how the relative positions of these key points change over time, the system defines and identifies different expressions.

(7) Robust Expression Recognition:

Through this method, the system detects and recognizes expressions by perceiving the relative positions of two or more key point coordinates within a single frame. This approach provides robust expression recognition and the ability to gather valuable classroom effect information.

In essence, this sophisticated facial expression model utilizes the power of data capture, real-time processing, and the rich information provided by MediaPipe Face Mesh to perceive, analyze, and understand facial expressions in real time. By monitoring the dynamic relationships between key facial points, it can accurately identify various expressions, making it a valuable tool for enhancing classroom engagement and facilitating effective teaching and learning (Fig. 4.2).

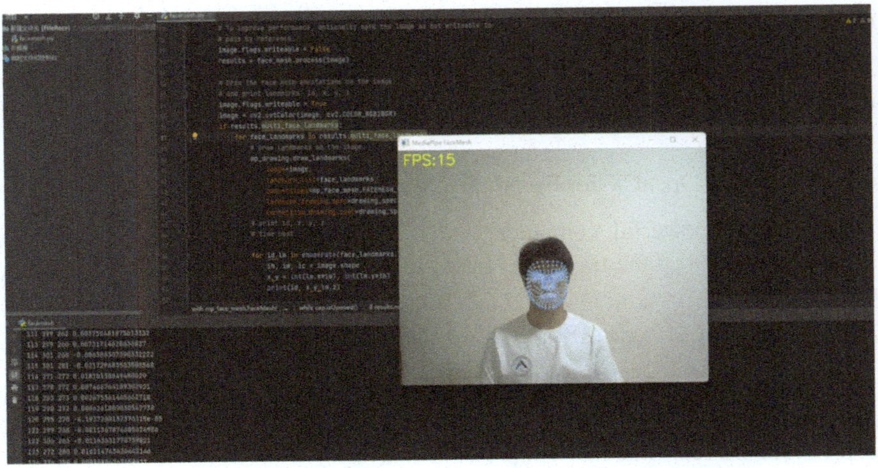

Fig. 4.2 Face detection effect of MediaPipe face mesh

4.2.2 Applying Face Detection Technology for Engaging Classroom Students

It is challenging to evaluate feelings because of their implicitness. At present, the primary methods are questionnaires, observation, and interviews. However, these evaluation methods are time-consuming and laborious, and most are summarized evaluation methods, separated from the scenario in the classroom. Therefore, to realize the demand of modern education to grasp students' situations in real time, the project uses artificial intelligence to collect students' real-time expressions in class and carry out feedback processing, which has the timeliness and reliability that the traditional method cannot surpass.

4.2.2.1 Index Selection

Through reading the existing literature, we have found that after the research on the mixed teaching mode and the influencing factors, we know that the evaluation of the learning effect refers to the process that the participants use the scientific method to evaluate the quality and level of the students with the goal of students' development. Therefore, based on the basic principles of science, operability, rationality, comprehension, and dynamics, and adding the principle of difference and comprehensiveness into actual operation, we divide the evaluation indicators of the learning state into the following categories (Table 4.1):

Considering that among the above five evaluation indexes, learning interest, learning attitude, and learning willpower are easily observed quickly, the other two

Table 4.1 Learning status evaluation index

Interpretation of indicators	
Learning interest	Emotional state in which a person has a positive perception of learning
Study attitude	Affirmative or negative tendency of a learner to learn more consistently
Learning motivation	Tendency to initiate and sustain a student's learning behavior and to direct it to a goal
Learning willpower	Ability to continuously overcome difficulties to achieve a learning goal
Learning values	Recognition of a certain value or function of an objective person, thing, or objects

indexes are unsuitable for real-time analysis. Therefore, learning interest and learning willpower will be this project's main evaluation indicators.

Index Quantification

(1) **Classification of Classroom Expressions**

Most theories currently divide facial expressions into six categories: happiness, sadness, anger, disgust, fear, and surprise. After research, the most prominent expressions in the learning process include boredom, depression, confusion, engagement, and joy. It can be roughly divided into positive expressions and negative expressions. Positive expressions include engagement and joy; negative expressions include boredom, depression, and confusion. The expressions expressed above are as follows (Table 4.2):

The emotions are dynamic. It can be seen that when the student is interested or attracted to the content, the student will appear engaged; when the problem is solved finally, the student will show joy and obtain satisfaction; similarly, when the student is not interested in the content of the course, his face will be bored; when the content of the course is challenging to understand, the student will become confused; when the learning process fails repeatedly, students will be depressed. Otherwise, when a student's facial feature is not apparent, it is defined as calm.

(2) **Learning Interests**

Learning interest refers to a person's positive cognitive tendency and emotional state of learning. Generally speaking, learning interests can be divided into direct and indirect learning interests. The former is directly caused by the material learned or the learning activity, the learning process itself. The latter results from learning activities. According to the existing data, when students have learning interests, they will be more involved in the classroom content and show expression of engagement and joy. Therefore, we know that the learning interest is directly proportional to the learning input degree. Consequently, we take the learning investment degree as the primary criterion for judging the learning interest. The learning interest is proportional to the proportion of input duration = (input duration and joy duration)/total time.

Table 4.2 Specific expression

Facial expression	Eyebrow	Eyes	Lips
Engagement	Overall rise, wrinkles may appear on the forehead	The upper and lower eyelids are separated, the eyeball is fixed in front for a long time, and the eyes are wide open	Less or no change
Happiness	Slightly upward on the outside, and crow's feet may appear	The lower eyelids of the eyes are slightly raised, and wrinkles may appear in the lower eyelids	The corners of the mouth move outward and upward, the lips elongate, and the lips may be separated to expose the teeth
Boredom	Slightly drooping	Eye angle slightly downward. Eyes may be slightly closed	Less or no change
Depression	Sagging on the outside with slight wrinkles on the forehead	Eyes down, eyes may be slightly closed or closed, eyes appear smaller	Close, lower lip bead up, mouth corner down
Confusion	Slightly drooping	The lower eyelids of the eyes are slightly drooping, and wrinkles may appear in the lower eyelids	Drooping corner of the mouth

(3) **Learning Willpower**

Learning willpower refers to an individual's ability to overcome difficulties and complete learning tasks continuously and is usually marked by the length of time the learner lasts for each learning activity. We find that some students lack perseverance and have poor self-control ability. When they encounter difficulties in their studies, they often refuse to think and retreat or turn to teachers and classmates for answers. They do not have permanence and stability in doing things, exit when they encounter minor problems, lack proper control over their behavior, are easily tempted by some external things, lack a sense of responsibility, and so on. Students with a strong will to learn can often study hard and for a long time. Therefore, it is easy to see that learning willpower is directly proportional to the duration of the learning activity.

(4) **Learning Attitude**

A learning attitude is a stable learning tendency, which refers to a learner's persistently positive or negative behavior tendency to study or ready condition of internal reaction. It can usually be judged and explained from the students' attention, emotion, and will learn. Students' learning attitudes can include the perspective of the course study, the learning materials, the teachers and the school, etc. Learning attitude regulates learning behavior, which is also reflected in students' reactions to the learning environment. Students work hard and actively when their perspective is consistent with the teaching environment. However, suppose there is a bad attitude to the learning environment (such as teachers and schools). In that case, they will

Table 4.3 Design of sequence diagram of student's personal expression change

2	
1	
0	
-1	
-2	
-3	
Time	0 5 10 15 20 25 30 35 40

Table 4.4 Design of sequence diagram of class's overall classroom expression distribution

1	
0.8	
0.6	
0.4	
0.2	
Time	0 5 10 15 20 25 30 35 40

avoid the learning environment and produce destructive behaviors, such as truancy and resistance. However, learning attitude depends on target interest, which combines the above two indicators. Therefore, we think that learning attitude is directly proportional to the enjoyment of learning activities (the proportion of joy and expression) and the degree of adherence to learning activities (the ratio of positive emotions).

Classroom Content Analysis Based on Evaluation Data of Learning Emotion

After collecting the expression data of all students, a graph can be drawn, such as a sequence diagram of expression distribution. Teachers and students can adjust the classroom based on the graph data and improve the outstanding problems (Table 4.3).

In the table above, 2 to −3 correspond to joy, engagement, calmness, boredom, confusion, depression, and the same below (Table 4.4).

In this table, the above-described learning state can be represented by different color blocks, such as the upper light color block representing a student presenting a positive emotion, the middle color block representing calmness, and the lower dark color block representing a student presenting a negative emotion.

The students' expression data and the learning state in class can be obtained through the above method, and an in-depth analysis is carried out according to the data. The corresponding reminder is sent to the system teacher's end, and the period data statistics and visualization are carried out in the system.

4.2.3 Face Detection Scheme Based on BlazeFace Detector

Facial expression recognition is a multi-classification problem with complex interference factors. Because the face has large and rich muscle groups, the human body's complex and diversified characteristic group determines the individual identity and expression. Currently, most theories classify facial expressions into six categories: happiness, sadness, anger, disgust, fear, and surprise. Most datasets in the existing are labeled according to the classification method. In this project, considering the super real-time performance and recognition accuracy of BlazeFace-based facial detectors, we use BlazeFace-based facial detectors.

The MediaPipe face detection scheme attempts to provide feedback on the current classroom learning status through expression recognition. For the classroom teaching environment, this project defines classroom expressions using the combination of expressions and eye movements (Fig. 4.3).

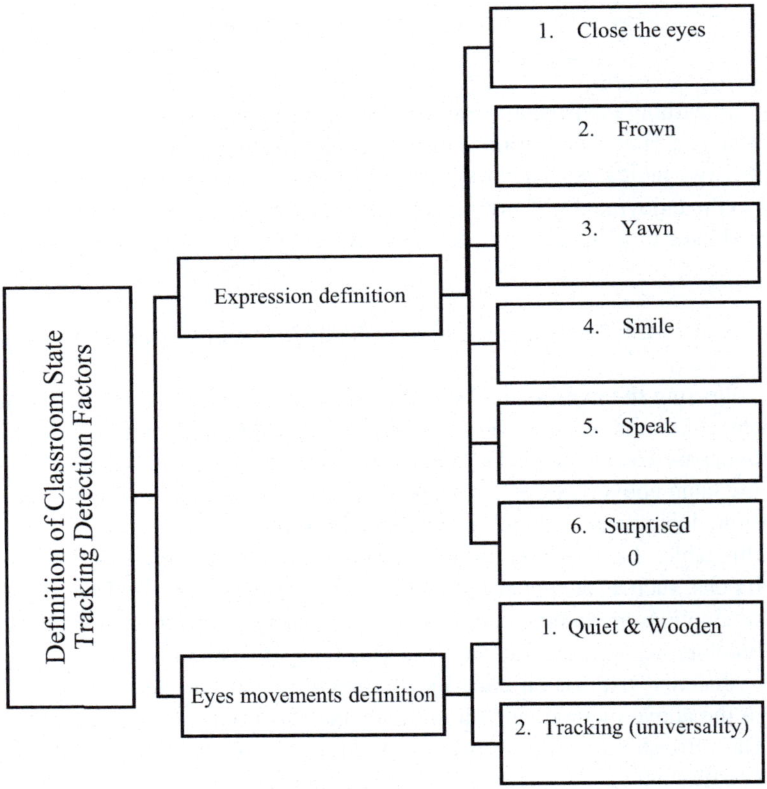

Fig. 4.3 Definition of classroom state tracking detection factors

Combined with the defining factors of classroom state tracking detection, it is integrated into two comprehensive dimensions of positive expression and negative expression:

(1) **Expression Recognition**

C. Positive expressions: 4, 5, 6
D. Negative expression: 1, 2, 3

(2) **Eyes Movements Recognition**

A. Positive expression: 2
B. Negative expression: 1

Track the classroom status using instant classroom behaviors, expressions, and eyes movements, and timely provide feedback on the current effect (e.g., teacher: difficult/easy, wonderful/plain; student: difficult/easy, accepted/unaccepted); accumulate information to evaluate teacher classroom effect (e.g., weekly, monthly, quarterly, annual class difficulty, excellent, and dull trend) and student learning effect (e.g., homework trend, interaction trend, and classroom performance trend); and provide instant (e. g., teacher: it is difficult, adjust it; student: You did not get it, do you? Do not miss it!) and periodic suggestions (e.g., teacher: the ratio of difficulty is 6: 4, students: more than 50% of homework has been submitted this week, and you should finish it in time!).

Due to the limitation of existing conditions, the project has not been fully implemented in this link, especially in optimizing behavior, expression, and eyes movement calculation model.

4.2.3.1 System Prototyping

The comprehensive project has been implemented, and we are trying to build a classroom application system based on dynamic face detection technology. The method comprises the following: Acquire face picture information of a student based on a classroom camera; judge the classroom state of the student based on the face picture information; input the face picture information into a trained image classification model, and consider whether the expression of the student is negative or not based on a vector output by the image classification model; when the negative expression of the student appears, send a corresponding reminder to a system teacher terminal, and time period data statistics and visualization are carried out in the system. Basic Design flowchart (Figs. 4.4, 4.5, 4.6 and 4.7):

Based on the design scheme of the system, we have preliminarily designed the operation interfaces of the client and the server for the system.

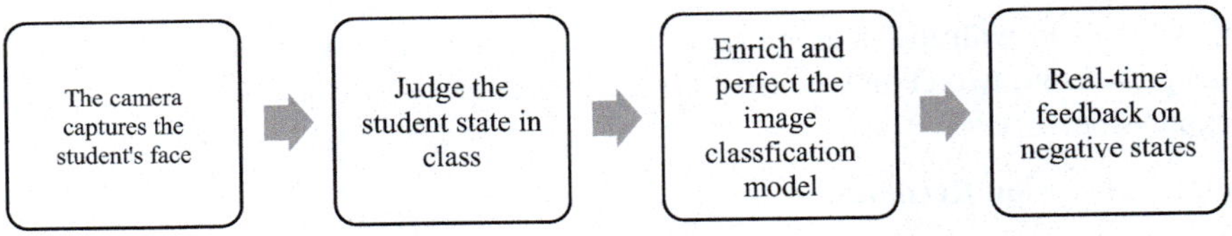

Fig. 4.4 Basic design flowchart

Fig. 4.5 Design of student client user interface

4.2.4 Personal Data Processing in Experiments

In recent years, with the rapid development of information technology, face detection has gradually penetrated all aspects of people's lives. The application of face detection can be seen from the smart city construction to the mobile phone clients' log-in and unlock-in. Face detection technology is important in many fields, including frontier defense, public transportation, urban security, and epidemic prevention and control. However, human face information, which is the biometric information in the sensitive personal information, is personal information with the most vital social attribute and the easiest to collect. It is unique and unchangeable. Once it is disclosed or illegally used, it will cause significant harm to personal and property safety. Therefore, personal information protection concerns the public's vital interests and the

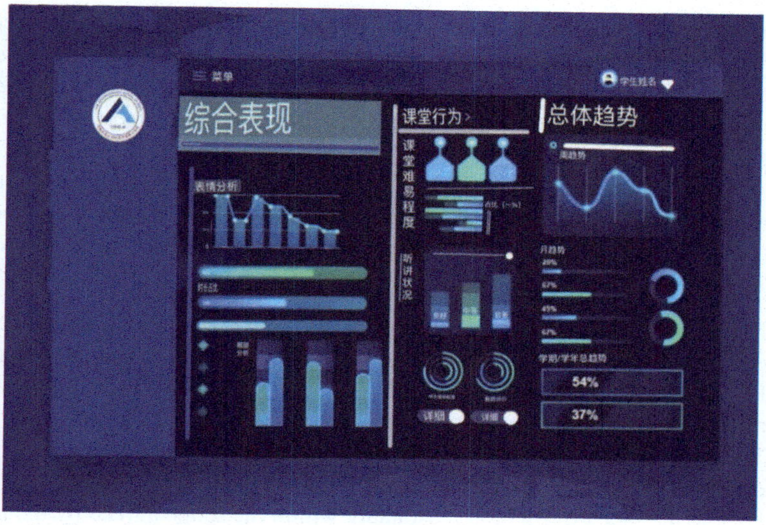

Fig. 4.6 Sever-side interface design

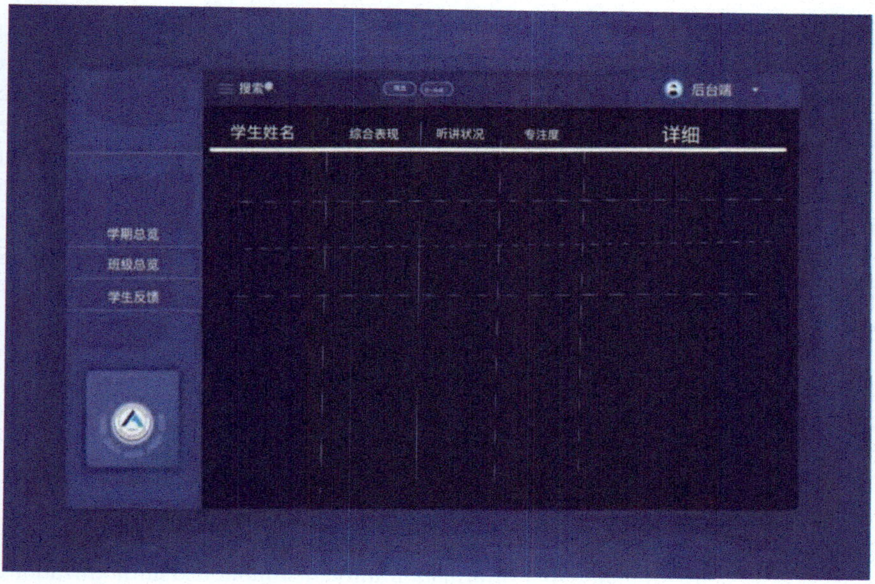

Fig. 4.7 Client user interface design for teacher

digital economy's healthy development. To ensure the excellent use and security of this advanced technology, the Supreme People's Court of China issued *Regulations of the Supreme People's Court on Several Issues Concerning the Application of Law in the Trial of Civil Cases Related to Handling Personal Information by Using Face Detection Technology* on July 28, 2021 (Hereinafter referred to as the *provisions*). Therefore, the use of face detection technology is restricted by law. In the provisions, it is specified that in any of the following circumstances, the people's court shall support the information processor's claim that he/she shall not bear civil liability:

(1) Dealing with human face information in response to a public health emergency or an emergency to protect natural persons' life health and property safety.
(2) Face detection technology is used in public places according to relevant state regulations to maintain public safety.
(3) To handle face information reasonably by carrying out news reporting, public opinion supervision, and other behaviors for the public interest.
(4) Handle the face information reasonably within the scope agreed by the natural person or his guardian.
(5) Other circumstances in conformity with the provisions of laws and administrative regulations.

The utilization of dynamic face detection technology in this project represents a significant stride toward enhancing the classroom learning environment through the process of facial expression recognition. This innovative approach seeks to provide educators with invaluable insights into the real-time state of students' engagement and comprehension. By promptly conveying information about students' learning statuses to teachers, parents, or school administrators, this technology empowers educators to make informed decisions and adapt their teaching strategies during the learning process, ultimately leading to an improvement in classroom teaching efficiency.

In line with the legal framework provided by Provisions (4), the project incorporates robust measures to ensure the responsible and ethical handling of facial information. These measures go beyond mere compliance and are designed to safeguard the privacy and security of students while delivering the intended educational benefits. One crucial aspect of this approach is the thorough analysis and processing of collected facial data on a backend server.

Instead of directly displaying images, the system employs a more discreet approach by conveying essential information through simple textual notifications. These notifications serve as alerts to teachers regarding specific conditions observed in the classroom. It is important to emphasize that these security and confidentiality measures are non-negotiable and are applied rigorously to protect the privacy of individuals.

Moreover, the project places significant emphasis on obtaining informed consent from both students' parents and educational authorities before implementing this technology. This consent process ensures transparency and clarity in how data is collected, stored, utilized, and eventually erased. Students and their parents are provided with clear explanations about the data lifecycle and are informed of their rights and how they can exercise control over their personal information.

The ultimate goal of this project is to strike a delicate balance between enhancing classroom efficiency and respecting individuals' personal privacy rights. By adhering to these ethical principles and legal guidelines, the project not only contributes to the educational advancement of students but also sets a precedent for responsible and privacy-conscious deployment of technology in the educational sector.

Furthermore, the implications of this technology extend beyond the classroom. It holds the potential for broader applications in the fields of public administration, security, health care, and beyond. For instance, in the realm of virtual meetings, the technology can be employed for personal identification and enhanced security. In health care, it can facilitate real-time monitoring of infants' well-being, providing peace of mind to parents and caregivers. Additionally, in crowded environments such as customs, airports, stations, and shopping malls, the technology can be harnessed to identify and respond to abnormal situations, thereby enhancing public safety.

4.3 Conclusion and Future Plans

The journey of our project has been a remarkable exploration into the dynamic realm of face detection technology within the educational landscape. This endeavor is not merely an academic pursuit but a passionate quest to reshape the very foundations of how we perceive and enhance the learning process.

At its core, the project sets out to establish a comprehensive, real-time monitoring, and control system that delves deep into the intricacies of students' learning states within the classroom environment. By harnessing the power of cutting-edge facial recognition technology, we aim to provide educators, parents, and school administrators with invaluable insights into the dynamics of classroom learning. This, in turn, equips teachers with the tools they need to adapt their teaching strategies in real time, fostering an educational environment that is not only more efficient but profoundly engaging.

Yet, the impact of this project transcends the confines of the classroom. It aspires to address long-standing challenges that have plagued education for generations. From classroom attentiveness to the retention of critical knowledge, these age-old issues have remained elusive, resisting conventional solutions. Our project's dynamic feedback loop holds the potential to be the catalyst for change, offering students personalized and responsive teaching approaches that promise a more enriching and effective learning experience.

As the project journey unfolded, our team embarked on a voyage of discovery and innovation. We immersed ourselves in a vast sea of research materials and leveraged our personal classroom experiences to shape the project's trajectory. The evolution of this endeavor was marked by a commitment to pushing boundaries, questioning norms, and boldly venturing into uncharted territory.

Our team's collective potential was brought to the forefront as we navigated the complexities of this project. Each member of our team assumed a unique role, capitalizing on their individual knowledge base and specialized expertise. Zhao Jiaming's mastery of face detection technology ensured the accurate identification of facial features, setting the foundation for our project's success. Chen Jiaxuan's expertise in facial expression recognition added an emotive dimension to our endeavor, allowing us to delve into the nuances of emotional interpretation. Deng Yuxuan, with her creative flair, crafted a visually captivating prototype that brought our project to life. This harmonious synergy among team members not only accelerated our progress but also fostered an environment of shared learning and growth.

While our project has made significant strides, we remain ever cognizant of its limitations. Practical testing within real-world classrooms has yielded valuable insights, yet certain areas demand further exploration and refinement. We recognize that the accumulation of more extensive and diverse datasets is paramount to enhancing the robustness of our research outcomes and validating the efficacy of our project beyond a shadow of doubt.

The future is a canvas awaiting our innovation and dedication. Our project's trajectory is clear—it is an unwavering commitment to continuous improvement and optimization. This ongoing journey encompasses not only the refinement of our existing face detection algorithm but also the fortification of our system's capabilities. Furthermore, we are dedicated to ensuring that the practical application of our technology aligns seamlessly with our project's founding principles and objectives.

As we venture further into the uncharted territory of dynamic face detection technology, we do so with a sense of responsibility. We are steadfast in our commitment to upholding the ethical and responsible use of this technology, ensuring that it serves as a force for good in the educational landscape.

In conclusion, our project is a testament to the boundless potential of human innovation and technology. It is a journey that transcends the boundaries of the classroom, impacting the lives of students, educators, and administrators alike. We stand at the precipice of a new era in education, one where the fusion of dynamic face detection technology and pedagogical expertise promises to redefine the art of teaching and learning. With each step forward, we inch closer to a future where education is not only efficient but also profoundly transformative.

Bibliography

Chen, T. (2020). Application of dynamic face recognition technology in university management. Electronic Commerce. *Electronic Commerce*, *11*, 93–94. https://doi.org/10.14011/j.cnki.dzsw.2020.11.040

Du, X., Huang, X., & Liu, Y. (2021). Overview of face recognition technology based on Python. *Electronic Testing*, *16*, 80–81+48. https://doi.org/10.16520/j.cnki.1000-8519.2021.16.032

Liang, X., Peng, R., Liu, H., & Zhang, Y. (2022). Intelligent teaching assistant platform based on face recognition. *Technology and Innovation Application*, *12*(6), 58–62. https://doi.org/10.19981/j.CN23-1581/G3.2022.06.017

Liu, C. (2020). *Dynamic face recognition based on deep learning* [Master's thesis, Southwest University of Science and Technology]. https://doi.org/10.27415/d.cnki.gxngc.2020.000029

Mu, T. (2019). *Research and application of face recognition technology based on deep learning* [Master's thesis, Xi'an University of Posts and Telecommunications]. https://kns.cnki.net/kcms2/article/abstract?v=3uoqIhG8C475KOm_zrgu4lQARvep2SAkOsSuGHvNoCRcTRpJSuXuqX-Bj5U5_-7r8yx8jfxAxZ-fJf6l1dAvYnvL2vXFDfSQ&uniplatform=NZKPT

Pan, X. (2022). Research on online learning effectiveness monitoring based on expression recognition. *Microcomputer Applications, 38*(3), 55–57.

Xu, M., & Zhang, R. (2018). Application of face recognition-based management system in student classroom. *Audio Engineering, 42*(9), 6–7+32. https://doi.org/10.16311/j.audioe.2018.09.003

Zhang, Y. (2021). *Key technologies research on multi-pose face recognition* [Master's thesis, Sichuan University]. https://doi.org/10.27342/d.cnki.gscdu.2021.000742

Lan Wu Information technology teaching and research team leader of the Second Branch School Attached to Renmin University of China, the academic leader of Haidian District, and the information technology teaching and research trainee. Wu was awarded as the national, Beijing and Haidian District excellent instructor of information technology.

Ye Gao Information technology teacher of the Second Branch of Renmin University High School Affiliated to Renmin University of China. Gao is the backbone of information technology discipline in Haidian District and a trainee teaching and research member of IT discipline in Haidian District.

Chapter 5
A Smart Monitoring System on Refrigeration with Neural Network Models

Pou Weng Huang, Xin Yue Li, and Kin Un Lam

Abstract This project utilizes technology to combat food waste and safety issues. We developed a smart monitoring system on the refrigerator that reminds users to consume the food before it expires, reducing household food waste. We trained a Faster RCNN model to identify food and developed a front-end page for user interaction. Although this smart system on refrigerators may not significantly reduce food waste, it aims to increase public awareness and inspire individuals to act. Furthermore, this technology can be optimized and extended to various industries, such as social security and elderly services, to enhance life quality and decrease food waste, a prominent global development issue. The project aims to reduce food waste and improve food safety through technology, promoting a more sustainable future.

Keywords Food waste · Pre-trained image model · Real-time detection

Highlights

This project presents an innovative solution to combat food waste and improve food safety through technology. The smart monitoring system on refrigeration with neural network models is a promising step toward a more sustainable future.

Pui Ching Middle School, Macau is a school that combines the traditional rigorous academic attitude with the spirit of free and open inquiry. Pui Ching is dedicated to innovations in education and the holistic development of students. Through constantly improving our school facilities and devoting resources to teacher development, the school aims to help students develop intellectually, mentally, physically, emotionally, and socially.

P. W. Huang · X. Y. Li · K. U. Lam (✉)
Pui Ching Middle School, Macau, China
e-mail: LamKinUn@gmail.com

P. W. Huang
e-mail: 2057122-4@g.puiching.edu.mo

X. Y. Li
e-mail: 2057119-4@g.puiching.eu.mo

R. Huang et al. (eds.), *A Collection of AI Innovations by Chinese Teenagers*, Smart Computing and Intelligence, https://doi.org/10.1007/978-981-97-3316-3_5

Challenges: The goal of this plan is to reduce household food waste and improve food safety using technological solutions.

Solution: Developed a refrigerator camera capable of detecting food items. Constructed a comprehensive database of food expiration dates, allowing the system to keep track of each item's storage date. Integrated the system with a website to provide users with timely reminders about approaching expiration dates, encouraging prompt consumption.

Outcome: Successfully developed and produced a prototype set of refrigerator products. Created a website system capable of reading and processing real-time camera data.

How are the smart monitoring system on refrigeration work?

The smart monitoring system on refrigeration utilizes technology to combat food waste and safety issues. It reminds users to consume the food before it expires, reducing household food waste. The system uses a Faster RCNN model to identify food and a front-end page for user interaction. Additionally, the required program is imported into the camera in advance and provided for a user to use the program at anytime and anywhere. The camera can be used on multiple refrigerators, which means the user can use the intelligent refrigerator assistant to be applied to any refrigerator. A neural network deep learning hypertext language technology platform is connected to realize backend and front-end operations to provide help for a better visual effect for the user.

How does the neural network models trained and applied

The neural network model used in this project is a Faster RCNN model, which is a type of pre-trained image model. The model is trained using a large dataset of food images to identify different types of food. The training process involves several attempts to train different models, and the results are improved by adding more pictures to the dataset. During the training process, the computer sometimes confuses specific items with similar items, but this phenomenon is gradually reduced by taking pictures of designated items and similar items together and informing the computer of the differences between them. Finally, the stride size is reduced in the training to allow the computer to train more carefully and present a perfect result.

Once the model is trained, it is applied to the smart monitoring system on the refrigerator. The system uses a camera to capture images of the food inside the refrigerator, and the images are processed by the pre-trained model to identify the food. The system then reminds the user to consume the food before it expires, reducing household food waste. The system also includes a front-end page for user interaction, allowing the user to view the status of the food in the refrigerator and receive reminders.

Can this technology be applied to other industries besides household refrigeration?

This technology can be optimized and extended to various industries besides household refrigeration. It mentions that this technology can be applied to other industries such as social security and elderly services to enhance life quality and decrease food waste, which is a prominent global development issue. The technology can be used to monitor the freshness of food in various settings, such as restaurants, supermarkets, and food processing plants, to reduce food waste and improve food safety.

5.1 Project Overview of Smart Monitoring on Refrigeration

5.1.1 Background

This project's noble mission to reduce food waste and enhance food safety addresses a pervasive issue that many households encounter regularly. The sight of forgotten canned goods at the back of the refrigerator or the discovery of once-fresh produce now transformed into a rotten and odorous mess is a common experience. Regrettably, this scenario often culminates in the disposal of perfectly good food into the garbage bin. The consequences of such actions are staggering: when each household discards even a single rotten apple, collectively, we waste an astonishing 7.7 billion apples. These discarded apples could have potentially provided sustenance to low-income individuals in need. It is this alarming phenomenon of food waste that motivated the young minds behind this project to drive change.

In an age dominated by technology, the logical step forward is to harness its potential to address this issue. Food waste, also known as food loss, encompasses the squandering of food resources at various stages of production, transportation, sale, and consumption due to a multitude of factors, including mismanagement and consumer habits. It is a global predicament, one that has garnered significant attention and concern.

Alberto, the Deputy Director of Economic and Social Development within the Food and Nutrition Division of the Food and Agricultural Organization (FAO), aptly described food waste as a global crisis during a press conference. This crisis has grown to international proportions, as evidenced by the United Nations Environment Programme's (UNEP) Food Waste Index report, published on March 4, 2021. The report's shocking revelation that nearly 1 billion tons of food are wasted worldwide each year represents a sobering reality—one that is roughly twice the previous best estimates.

In this context, the significance of projects like this one cannot be overstated. Leveraging technology to mitigate food waste is not merely an innovative endeavor; it is a vital step toward a more sustainable and equitable world. Through the dedicated efforts of these young individuals, we glimpse a future where the impact of food waste is drastically reduced, making strides toward a world where food security is more attainable for all (Fig. 5.1).

The issue of food waste is one that resonates deeply with many individuals, particularly in today's age of technology and heightened awareness of sustainability. Have you ever opened your refrigerator at home and discovered long-forgotten cans with

Fig. 5.1 Ranking of food waste regions in the world

sedimentation? Or perhaps you have reached for an apple, only to find it rotten and emitting a foul odor? It is a common experience—spoiled food being discarded into the trash bin. But have you ever stopped to consider the collective impact of such actions?

Imagine a scenario in which every household around the world discards one rotten apple from their refrigerator. The sheer scale of this waste would equate to approximately 7.7 billion good apples—a quantity that could potentially feed millions of low-income individuals. It is a stark representation of the food waste phenomenon that teenagers are increasingly determined to change. In this era of advanced technology, there is an opportunity to leverage these innovations to tackle this pressing issue.

Food waste, or food loss, encompasses the disposal or wastage of food during various stages of its lifecycle, including production, transportation, sale, and consumption. This wastage occurs for a variety of reasons, from mismanagement to consumer habits. It is a global challenge that has garnered significant attention, as highlighted by Alberto, Deputy Director of Economic and Social Development at the Food and Agricultural Organization (FAO). Food waste has transitioned from being a localized concern to an international issue of paramount importance.

A report by the United Nations Environment Programme (UNEP) revealed staggering statistics. Globally, nearly 1 billion tons of food go to waste each year—a figure that is approximately double previous estimates. This waste is most pronounced in households, accounting for a substantial 11% of all food available to consumers. Food service and retail sectors contribute 5% and 2%, respectively, to this alarming statistic. At the individual level, the data is equally concerning, with an average of 121 kg of food wasted per capita, of which 74 kg are attributed to households.

Food safety is another critical facet of this issue. Ensuring that the food we consume is safe and free from contaminants is vital for our well-being. Foodborne diseases, often caused by bacteria, viruses, parasites, or chemical substances present in contaminated food or water, pose a significant threat. This is where the concept of food shelf life comes into play—it is a key determinant in maintaining food safety.

The entire process of food production, storage, handling, and consumption plays a pivotal role in food safety. Governments, international organizations, scientists, the private sector, and civil society all have crucial roles to play in ensuring food safety. This involves adhering to global food standards, establishing robust food regulatory systems with emergency preparedness measures, providing access to clean water, implementing good agricultural practices across various sectors, strengthening food safety management systems employed by food operators, and educating consumers to make informed choices about their food.

In light of these challenges, the innovation introduced by this project carries significant weight. By leveraging technology to address household food waste and enhance food safety, it offers a compelling solution to these pressing issues. The project's impact extends far beyond the realm of waste reduction; it contributes to a more sustainable and secure food supply chain for all, with potential benefits spanning from individual households to global communities. As the project progresses, it holds the promise of driving positive change in the way we perceive, manage, and consume food, ultimately shaping a more sustainable and secure future for all.

5.1.2 Project Impacts and Benefits

Social Impacts

(1) **Enhancing Environmental Sustainability and Health**: The primary objective of this project is to leverage technology for reducing food waste in households, which subsequently contributes to improved environmental sustainability. By

providing regular reminders to users about the optimal consumption period of food items, the project aims to prevent the unnecessary disposal of expired food. This approach aligns with global efforts to reduce food waste and supports a healthier lifestyle by ensuring that individuals consume safe and fresh food. Furthermore, the project promotes environmental awareness and responsible consumption by offering reports and articles on food waste and safety through its website.

(2) **Advancing Research in Social Science and Technology**: The project represents a novel intersection of social science and technology research. It addresses complex societal issues from multiple perspectives and employs innovative technological solutions. By developing a portable camera with pre-installed software, users can conveniently access this solution across various refrigerators, making it a versatile tool for mitigating food waste and promoting food safety. Additionally, the integration of neural network deep learning technology enhances the project's user experience, aligning with contemporary trends in technology and user interface design.

Economic Benefits

(1) **Mitigating Food Waste and Maximizing Safety**: One of the project's significant economic benefits is its capacity to reduce food waste effectively. By regularly reminding users about food preservation guidelines, it minimizes the disposal of expired food items. This, in turn, contributes to economic savings for households while ensuring food safety.

(2) **Market Attractiveness and Accessibility**: The project's unique features, convenience, and affordability position it as an attractive solution for a diverse user base. Its accessibility is akin to purchasing an online camera with embedded software, making it suitable for a broad range of users. This project appeals to individuals across age groups, from busy young professionals seeking convenience and eco-consciousness to elderly individuals who may benefit from timely reminders. Its alignment with real-life needs and its ability to address critical societal challenges make it poised to capture a significant market share and generate economic value.

Generalizability

The project's overarching goal is to reduce food waste, enhance the quality of life, ensure food safety, promote environmental conservation, and foster healthy living through the application of science and technology. Its versatility is evident across several dimensions:

(1) **Societal Impact**: The project's portability and adaptability across different environments position it as a tool for reducing food waste at the household level, ultimately contributing to a more environmentally conscious and health-focused society.

(2) **Economic Accessibility**: Its adaptability to a wide range of user demographics and innovative approach ensure its long-term viability as a convenient technological solution for modern living.
(3) **Technological Integration**: By leveraging neural network technology for website connectivity and designing a user-friendly front-end, the project remains attuned to societal needs, facilitating straightforward and effective utilization.

In summary, this project transcends its immediate objectives related to food waste and safety, extending its potential impact to encompass broader societal and economic contexts. Its adaptability and innovative approach position it as a valuable tool for promoting responsible consumption, environmental awareness, and overall well-being.

5.2 Project Design for Modifying the Refrigerator

5.2.1 Schemes and Timetable

Scheme 1: Place a camera in the refrigerator. A miniature weighing instrument is arranged in the fridge, which is matched with the camera to detect the freshness of the food. And then, the mobile phone software is used to remind the expiration of the shelf life

Scheme 2: Make a camera in the refrigerator that can detect food. Moreover, there is an extensive database of shelf life and will be classified food. When the shelf life is approaching, combine it with the website to remind users to eat as soon as possible. It is guaranteed to reduce unnecessary household food waste and ensure food safety

Regarding the first scheme, placing the camera in the refrigerator is not practical, and there is no suitable miniature scale to place. In addition, detecting the freshness of the food with the camera requires a vast database and some olfactory processing. So, we chose scheme 2, which is more logically and technically feasible, as the initial scenario.

Scheme 1 and Scheme 2 represent two distinct approaches to address the issue of food waste and safety through technological intervention in household refrigerators. While both schemes aim to reduce food waste and ensure food safety, Scheme 2 emerges as the more practical and technically feasible choice for the following reasons:

Scheme 1: Camera with Miniature Weighing Instrument

Practicality and Space Constraints: Placing a camera and a miniature weighing instrument in a refrigerator presents practical challenges. Refrigerators typically have limited space, and adding such equipment may interfere with the storage of food items and reduce the refrigerator's capacity.

Complexity of Freshness Detection: Detecting food freshness solely through a camera and miniature weighing instrument may be limited in its accuracy. Assessing food freshness often requires sensory cues such as smell, texture, and color, which are challenging to replicate with this setup.

Database and Olfactory Processing: Building a comprehensive database for food freshness and integrating olfactory processing capabilities into the system would be highly complex and resource-intensive.

Scheme 2: Comprehensive Camera-Based System

Food Detection and Classification: Scheme 2 proposes a camera-based system that can detect and classify food items within the refrigerator. This approach allows for a more accurate assessment of food freshness based on visual cues.

Extensive Database: Scheme 2 leverages an extensive database of shelf life information for various food items. This database enhances the accuracy of freshness assessments and enables precise expiration date reminders.

Practical Implementation: Scheme 2's approach is more practical and user-friendly. Users can easily interact with the system through a mobile app or website, receiving timely reminders about food items nearing their expiration dates.

User Convenience: In Scheme 2, users benefit from a more user-friendly and convenient experience. The system's ability to classify food items and provide accurate freshness information simplifies the user's role in managing their refrigerator's contents.

Integration with Existing Technology: Scheme 2 can seamlessly integrate with existing technologies, such as smartphones and the internet, making it accessible to a wide range of users. This integration ensures that users can receive reminders and updates conveniently.

Food Safety Assurance: The extensive database and accurate freshness assessments in Scheme 2 enhance food safety by reducing the chances of consuming expired or unsafe food items.

Given these considerations, Scheme 2 aligns more logically and technically with the goal of reducing household food waste and ensuring food safety. It offers a feasible and user-friendly solution that integrates well with modern kitchen environments and consumer habits. As technology continues to advance, Scheme 2 holds the potential to make a significant impact on reducing food waste and improving food safety practices in households around the world (Table 5.1).

Table 5.1 Timetable of the project

October–November 2021	Preliminary plan conception
November–December 2021	Collecting images
January–March 2022	Neural networks and website development
April–June 2022	Testing and adjusting
July–September 2022	Self-study (knowledge accumulation)
September–October 2022	Plan optimization
After October 2022, Add the database and update the website configuration	

5.2.2 *Breakthrough in Difficulty*

5.2.2.1 How to Install the Camera on the Refrigerator

The evolution of the solution from the initial implementation of Scheme 2 to the final design with a camera on top of the refrigerator highlights the importance of creative problem-solving and practical considerations in technology development. Here's how the process unfolded:

Initial Challenges

(1) **Cryogenic-Resistant Camera**: The initial concept involved placing a cryogenic-resistant camera inside the refrigerator. However, this approach presented several challenges, including the high cost of such cameras and the need for multiple cameras to cover different compartments, which could drive up the overall price significantly.
(2) **External Box**: The second idea was to create an external box outside the refrigerator with a camera mounted on top of it. This setup aimed to detect items and interact with a web address. However, this approach encountered limitations in capturing the entry and exit status of food items effectively.

Refinement of the Solution

After careful consideration and two rounds of discussions, the project team arrived at a more practical and effective solution:

Camera Placement: Placing a camera on top of the refrigerator to detect food varieties and monitor the entry and exit status of items. This approach offered several advantages:

Cost-Effective: It eliminated the need for cryogenic-resistant cameras and complex external boxes, making the solution more cost-effective and accessible to users.

Comprehensive Detection: The camera on top of the refrigerator could capture a wide view of the refrigerator's contents, enabling it to detect various food items effectively.

Entry and Exit Monitoring: By strategically positioning the camera, the system could track the entry and exit status of food items, providing users with accurate information about their refrigerator's contents.

User-Friendly: This solution aligned with user expectations and convenience, as it did not require significant modifications to the refrigerator's interior or additional equipment.

Simplicity: Placing the camera on top of the refrigerator simplified the implementation process, ensuring that the solution could be easily adopted.

Taking a New Pace

With the decision to place the camera on top of the refrigerator, the project made significant progress and gained momentum. This refined approach combined practicality and cost-effectiveness, addressing the initial challenges and ensuring that the system could effectively reduce food waste and enhance food safety.

Innovation often involves creative problem-solving and adaptability, as seen in the evolution of this project's solution. By continuously evaluating and refining ideas, the project team arrived at a solution that holds promise in addressing the critical issue of food waste and safety in households. This process serves as a valuable example of how technology development can benefit from thoughtful consideration and iterative problem-solving.

5.2.2.2 How to Detect the Entry and Exit Status Items

The process of refining the system's design from using hand tracking to object tracking and considering the camera's placement and angle adjustment demonstrates the project team's dedication to finding the most practical and efficient solution. Here's how the project's design evolved:

Initial Idea: Hand Tracking

The initial concept involved using hand tracking as a means to determine the trajectory of objects entering and exiting the refrigerator. However, this approach presented several challenges:

(1) **Heavy Workload**: Implementing hand tracking would require significant computational resources, adding to the system's workload.
(2) **Additional Neural Network**: Hand tracking would necessitate the use of an additional neural network for deep learning, further complicating the system.
(3) **Recognition Errors**: The presence of the hand could obscure or interfere with the recognition of objects, potentially increasing recognition errors.

Revised Approach: Object Tracking

Recognizing the limitations of hand tracking, the project team shifted to a more efficient and accurate method: object tracking. In this approach, the system focused on tracking the objects themselves rather than the hands interacting with them. This change brought several advantages:

(1) **Reduced Workload**: Object tracking is computationally less intensive than hand tracking, reducing the system's workload.
(2) **Improved Accuracy**: By tracking the objects directly, the system could more accurately monitor their movements without the interference of occluding hands.

Camera Placement and Angle Adjustment

To ensure the effectiveness of object tracking, the team had to consider the camera's placement and angle:

(1) **Camera Position**: Placing the camera on top of the refrigerator provided a suitable vantage point for tracking the objects' movements.
(2) **Entry and Exit Patterns**: Observations revealed that objects were typically placed into the refrigerator from the bottom to the top and removed from the top to the bottom. This pattern informed the system's entry and exit status judgments.
(3) **Center Point Matching**: By matching the center point of the tracked object with its movement trajectory, the system could determine its entry or exit status accurately.

Achieving Entry and Exit State Detection

With the refined approach of object tracking and careful consideration of camera placement and angle, the project team successfully developed a system capable of detecting the entry and exit states of items in the refrigerator. This achievement marked a significant milestone in addressing the project's goals of reducing food waste and enhancing food safety.

The evolution of the system's design highlights the importance of adaptability and practicality in technology development. By selecting the most suitable methods and making necessary adjustments, the project team optimized the system's performance and usability, ultimately delivering a more effective solution to the challenge at hand.

5.2.2.3 How to Improve the Effect of Item Recognition

Initially trying to increase the number of image datasets, each product added 200 images to the data set. However, the effect is not ideal, and the error is still significant. Slowly we start trying to collect images from different angles of some products. For example, for an apple, we collect several images of it looking down, looking up, left, right, and 45°, etc. It is worked this way. Finally, since depth learning uses the convolution layer for recognition, the detection can be more careful and detailed by adjusting the recognition stride. The error is also reduced by reducing the stride of the filter (Fig. 5.2).

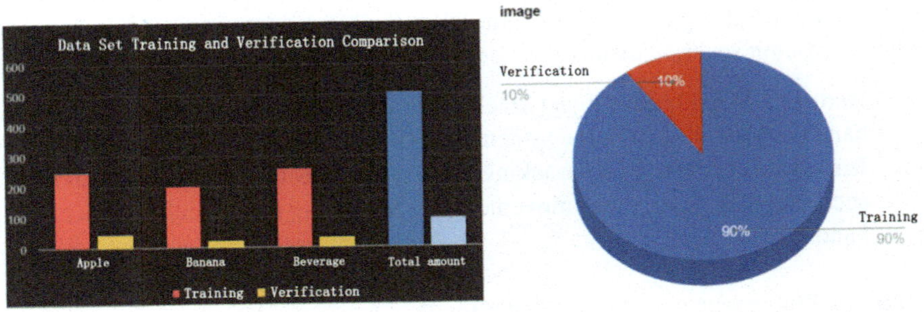

Fig. 5.2 Data set training and verification comparison

5.3 Implementation of Smart Monitoring System on Refrigeration

The primary function of the intelligent refrigerator assistant is to recognize the classification of the items and judge whether to put or take out the goods when the user puts them in front of the camera and records them on the website (Figs. 5.3, 5.4 and 5.5).

(1) The function of distinguishing access is achieved by recording the current time, judging, judging the items' motion track, and using the motion track of the center to deduce the user's action.

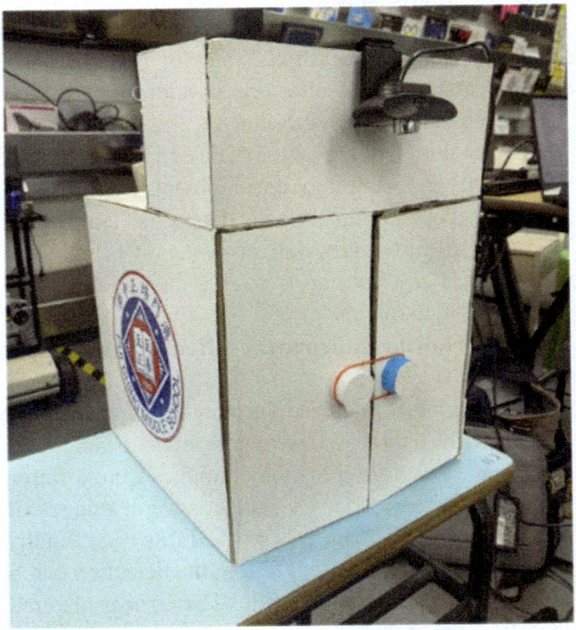

Fig. 5.3 Prototype of refrigerator

Fig. 5.4 Detection of articles

Fig. 5.5 Identifying user actions as putting in or taking out

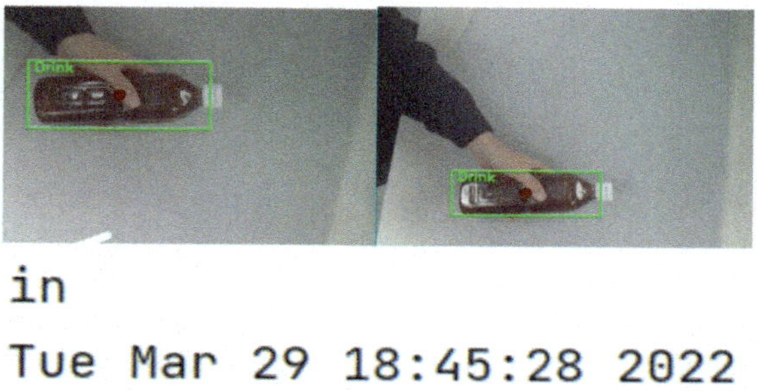

(2) When an object is detected, and the entry and exit direction of the article is judged to be put in storage, the user is indicated to put the item in the right direction. Otherwise, the user takes the item out of the refrigerator.
(3) When the entry time of any goods is collected, the time point will be stored on the website. The intelligent refrigerator assistant can take out the food which reaches the shelf life at the fastest time, judge the best use time of it, and remind the user when the user does not take out the article by the time. If the goods have expired, a display will inform the user that the item has expired.

This scheme provides an accessory device for the refrigerator. Considering that the project is not entirely improved at present, the camera of the accessory equipment is still connected to the computer and performs the operation. However, after the solution is perfected, the program and database stored in the computer will be held on the GPU to achieve the effect of portability and easy installation.

Fig. 5.6 Creating a dataset

5.3.1 Deep Learning Process

In terms of deep learning, the system is divided into four parts.

5.3.1.1 Creating Dataset

A network crawler was conducted to classify and name files of images on the Internet with the items required for the solution, put them into an archive named labeling, and mark the classification target of each image. Create the data set for the next model. Computers need much data to support them before they can learn in depth (Fig. 5.6).

5.3.1.2 Generating Model

In the model, this scheme uses the object-oriented method in PyTorch software. Object-oriented programming (OOP) is a function statement in Python that can create new objects. This project uses the class to define new objects for processing pictures and labels.

This project uses the function of transfer learning. Transfer learning is a research field of machine learning. It focuses on storing a solution model for an existing problem and using it on other different but related issues. The proportions have been pre-trained, so this scheme can achieve better results using a relatively small number of images. This project divides the collected 779 pictures into a training set and a verification set and re-learns it on the weight that the Master R-CNN had, which accelerates the neural network's learning process, and finally presents a deep learning result of the neural network.

This model defines three functions in the class: defining the initial value, calculating the number of images, dividing the pictures into a training set and a validation set, and processing photos and tags (Fig. 5.7).

The processing of the image and tag is divided into *x* and *y*. Image processing in the model mainly refers to the adjustment pattern, which converts the image data

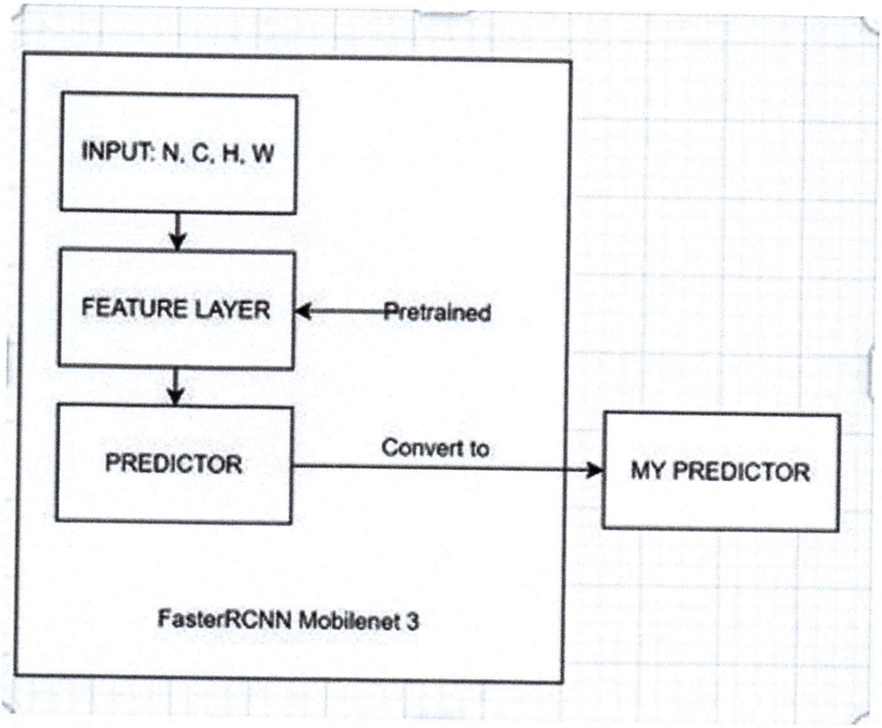

Fig. 5.7 Fast R-CNN prediction process

from H, W, C to C, H, W. Adjusted to a number between 0 and 1 and converted to PyTorch form.

Y (label) is used to identify the object's position on the picture and draw a box. How to get the coordinates of the box of the specified item:

(1) Goal: Find the top left and bottom right corners of the box
(2) Methods: Use map syntax to find the center point coordinates (cx, cy), height (h), and width (w) of the box. Find the top left corner and the bottom right corner using the formula. The *x*-coordinate on the top left corner: use the *x*-coordinate of the center point to subtract the length of the width and divide by two. The *y*-coordinate on the bottom right corner: use the *y*-coordinate of the center point to subtract the height and divide by two. The other lower right corner is the same, which changes the subtraction into the increase. The following is the practice diagram (Fig. 5.8).

5.3.1.3 Model Test

The selection of Faster R-CNN as the deep learning framework for this project marked a crucial step in achieving accurate object detection within the refrigerator.

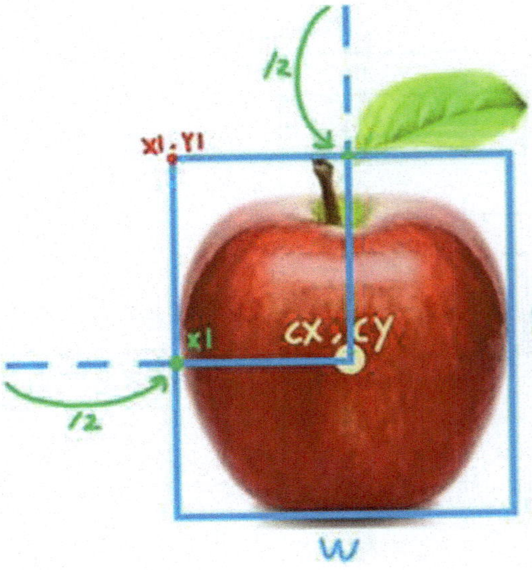

Fig. 5.8 Method of object positioning

Faster R-CNN, a type of regional convolutional neural network, features specific components that contribute to its effectiveness in two-dimensional data processing, particularly in image recognition. Here's an overview of the key components and the challenges faced during the training process:

Components of Faster R-CNN

(1) **Convolution Layers**: Faster R-CNN incorporates one or more convolution layers, which are fundamental for processing image data efficiently. These layers perform operations that enable the network to recognize patterns and features within the input images.
(2) **Fully Connected Layer**: At the top end of the network, there is a fully connected layer. This layer plays a crucial role in making predictions based on the features extracted by the preceding layers.
(3) **Correlation Weights**: These weights help establish connections between different parts of the neural network, enabling the model to consider the relationships between various features.
(4) **Pooling Layers**: Pooling layers are employed to downsample and reduce the dimensionality of the feature maps, which aids in improving the network's efficiency and performance.

Training Challenges and Solutions

(1) **Initial Training Failure**: The first attempt at training the model resulted in failure. The computer incorrectly identified all cuboid and round objects as drinks and apples. This failure highlighted the need for more extensive and diverse training data.

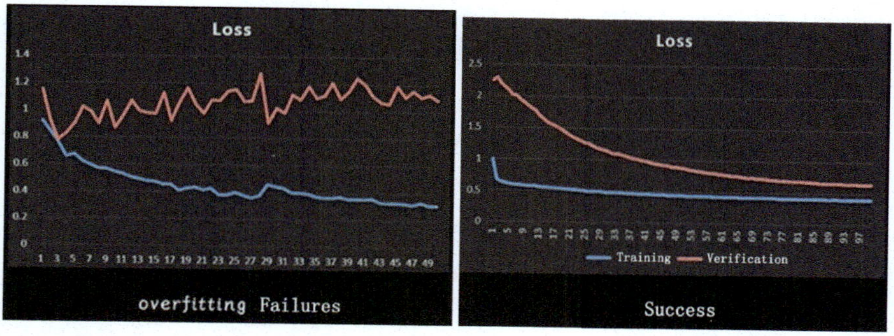

Fig. 5.9 Error diagram in training

(2) **Data Set Enhancement**: To address the initial failure, the team introduced additional images into the training data. This helped enhance the model's ability to distinguish between various objects within the refrigerator.
(3) **Confusion Between Similar Items**: During the training process, the computer sometimes confused specific items with similar objects. To mitigate this issue, the team collected images of designated items alongside similar items and provided the computer with clear distinctions between them.
(4) **Stride Size Reduction**: The stride size used in training was reduced to allow the computer to train more meticulously, resulting in improved detection accuracy.

The accompanying line chart illustrates the progression of training errors. The chart on the left represents the initial training error, while the chart on the right depicts the final training error for objects entering and exiting the refrigerator. This visual representation demonstrates the significant improvements achieved through iterative training and data augmentation.

Overall, the adoption of Faster R-CNN, coupled with diligent efforts to enhance the training dataset and address specific recognition challenges, led to the development of a more accurate and effective object detection system within the refrigerator (Fig. 5.9).

5.3.2 *User Interface Design*

The model obtained after the neural network training is taken out and applied to the real-time program.

(1) Draw the position and key points of the object. Use cv2 to take out the positions of the four sides of the object and use the rectangle syntax to frame the object. Recall the label corresponding to the object in the list of Coco classes, mark the object's name with cv2.putText, calculate the position of the middle point by using the position data of the four corners, and mark the center point with the circle.

Fig. 5.10 Item attributes

(2) Get the center point and judge the motion path when the object is detected. The position of the center point is calculated in the acquired image of each frame. Compare the change in the position of each central point. When the numerical value reaches a specific defined value, it will be considered in or out.

The program uses Flask to realize the item identification and website connection and does the last step that the program wants to complete—record the items stored in the user's refrigerator and remind them of the shelf life of the items (Fig. 5.10).

5.3.2.1 General Layout

The project utilizes hypertext markup language (HTML) and Flask to create a web-based platform. Upon logging in, users are presented with four main sections, each serving a distinct purpose:

(1) **Home**: This section likely provides an overview of the user's connected refrigerator and the items stored inside it. Users may interact with their refrigerators, view detected results, and receive real-time information about the freshness of their food.
(2) **Record**: The "Record" section probably keeps a history of the user's interactions with the intelligent refrigerator assistant. This could include timestamps of when items were placed in or removed from the refrigerator, helping users track their food management habits.
(3) **Total**: In this section, users may find aggregated data or statistics related to their food storage and consumption habits. It could provide insights into how effectively they are managing their food items.
(4) **Save**: "Save" serves as a platform for storing data on the website's backend. This could include information about the user's food catalog, access times, and entry and exit statuses of items. By sending this data to the website using methods

like labels and 'time.localtime()', the project ensures that users can access and manage their data from any device, such as their mobile phones or computers.

This platform aims to provide users with a comprehensive solution for food management, combining real-time monitoring, historical records, statistics, and data storage capabilities to reduce food waste and enhance food safety.

5.3.2.2 Homepage Design

The website's homepage appears to serve as an informative hub, offering various content and resources related to food waste reduction and food safety:

(1) **Articles**: The presence of articles on the homepage suggests that the website provides educational and informative content about food waste and food safety. These articles likely cover topics such as the impact of food waste, tips for reducing waste, and guidelines for ensuring food safety.
(2) **Advocacy**: The website advocates for the reduction of food waste, reflecting a commitment to raising awareness and promoting responsible food consumption. This advocacy may include campaigns, pledges, or calls to action aimed at encouraging visitors to actively participate in waste reduction efforts.
(3) **Introduction**: A general introduction to the program is available on the homepage. This section likely provides an overview of the project's goals, features, and benefits. It may also explain how the intelligent refrigerator assistant works and its potential impact on reducing food waste.
(4) **Video**: The presence of a video suggests that the website offers multimedia content to engage visitors. This video could provide a visual demonstration of the intelligent refrigerator assistant in action, showcasing its functionalities and benefits.

Overall, the homepage of the website appears to be a valuable resource for individuals interested in food waste reduction and food safety. It combines educational content, advocacy efforts, program details, and multimedia elements to inform and inspire visitors to take action against food waste (Fig. 5.11).

Fig. 5.11 Homepage design

5.3.2.3 Save Page Design

The Save platform on the website seems to serve as a data storage and retrieval system for the intelligent refrigerator assistant project. It stores essential information about items, including their type, access time, and access status. The process appears to involve the following steps:

(1) **Data Collection**: The intelligent refrigerator assistant collects data about items placed in and removed from the refrigerator. This data likely includes details such as the type of item (e.g., apple, milk, etc.), the time it was accessed, and whether it was put into or taken out of the refrigerator.
(2) **Data Storage**: The collected data is then transmitted to the Save platform on the website. This platform is responsible for receiving, organizing, and storing the data for future reference.
(3) **Data Format**: It appears that the data is stored in a text (txt) file format. Each piece of information, including the item type, access time, and access status, may be structured within the text file. This format allows for easy storage and retrieval of data.
(4) **Subsequent Recalls**: The stored data can be recalled and accessed through the Save platform when needed. Users or administrators may retrieve information about items' history, including when they were placed in the refrigerator and when they were removed.

By maintaining this database of item-related information, the Save platform contributes to the project's goal of reducing food waste and enhancing food safety. Users can monitor and manage their food items more effectively, thereby minimizing waste and ensuring the freshness and safety of the food they consume.

5.3.2.4 Record Page Design

It appears that the Save platform on the website has a feature that allows users to view their access records for items stored in the intelligent refrigerator assistant. Here's how this process works:

(1) **Data Retrieval**: The platform reads the data stored in the txt files, which includes information about the type of items, access times, and access statuses.
(2) **Data Presentation**: Once the data is retrieved, it is presented on the website's user interface. Users can access a dedicated section, such as the "Record" or "Access History" section, where they can view their access records.
(3) **Display Format**: The access records are likely displayed in a structured format on the website. Each record may include details such as the name of the item, the date and time it was accessed, and whether it was put into or taken out of the refrigerator.
(4) **User Accessibility**: Users can access and view their access records at any time through the website's interface. This feature allows users to keep track of when they placed items into the refrigerator and when they removed them.

Fig. 5.12 Record page design

(5) **User Benefits**: This access record feature provides transparency and accountability for the items stored in the refrigerator. Users can monitor their food items, track their usage, and ensure that they are consuming items within their optimal use periods.

Overall, this feature enhances the user experience and helps individuals make informed decisions about their food items to reduce waste and promote food safety (Fig. 5.12).

5.3.2.5 Total Page Design

In terms of total, the program counts the number of items of each different category and outputs the items with the fastest shelf life of each category on the website. If they have expired, we will use a pop-up warning box to warn the user that such items have expired. In terms of statistical quantity, the scheme adds up the last item of each item with the same label (i.e. 1 and -1 for in and out) so that the incoming statistics will be added by one, and the outgoing one will be -1, and the resulting quantity is the total quantity of each item. In terms of representing the fastest expired items, the scheme uses a dictionary syntax to divide each piece of data into three sequences from the category. In the three sequences, the scan starts from the next oldest record to offset each entry and exit (i.e. judging according to the 1 and -1 in the last column). Offset the In-and-out and eliminate data. Finally, look at the data in the dictionary to find the same item that has been put in for the longest time, and this item is the one we are looking for with the fastest expiration date.

5.3.3 Optimization

5.3.3.1 Disadvantages of the Current Scheme

(1) Object Differentiation: The current neural network struggles to accurately distinguish between different objects of the same variety. For instance, if two apples are placed in the refrigerator simultaneously, the system may have difficulty determining which one was put in first. The existing solution is to assume that

the user takes out the item that has been in the longest. However, this approach is not foolproof and may lead to inaccuracies in tracking freshness.

(2) Reminder Mechanism: The current reminder mechanism has room for improvement. Since the project is in a web version, it can only remind users when the web application is open. This limitation means that users may miss reminders when they are not actively using the application, potentially resulting in overlooked food items.

(3) Variety of Neural Networks: The dataset used for training the neural network is limited in terms of the variety of items it can accurately identify. With only a small dataset, the system may struggle to recognize a wide range of food items that users typically store in their refrigerators.

These challenges highlight areas where further development and refinement are needed to enhance the system's accuracy and user-friendliness. As the project progresses, addressing these issues will be crucial to achieving the goal of reducing food waste effectively and ensuring food safety.

5.3.3.2 Updatable Features

The portable version of mobile phone software is enhanced. When the user can download the mobile phone client, the app can promptly send out a warning of their food's expiration through the mobile phone promptly, thus improving the reminder mechanism.

When the database is multiplied, and the user is combined with the database picture enhancement, the user is willing to take photos, record some items which are not recorded or not recognized, carry out word supplements, and then store them in the database, which can make the user more satisfied with the system and increase the picture storage of the database. It is said that kill two birds with one stone.

Remind a user that the user needs to buy food materials when the user inputs food information and records the optimal service life of all the food. Calculate the frequency of the user using the refrigerator according to the state in which the user takes out the food. Combined the frequency of using the refrigerator with the time of the optimal use period to make a calculation and figure out how long the rest food in the refrigerator (excluding the canned food, which can be placed for a long time but is not frequently used) in the refrigerator needs to be eaten by the user and the time when the refrigerator is about to be cleared to remind the user of the time required to go shopping.

The matched nutritious food is identified and entered through object detection according to the food the user places. In the procedure, add a matching link. Add each food's basic protein quality, fat content, nutritional value, etc., to the database to match the healthy diet menu. Tips on eating healthy can be attached to remind the user of the importance of healthy eating.

According to the user's weight, health condition, allergy condition, etc., determine their healthy diet.

5.4 Conclusions

One-third of the food produced globally is discarded each year. That is, about 1.3 billion tons are dumped or wasted. Food waste includes food discarded during production, storage and transportation, processing and packaging, buying, and consumption. So, food waste in cooking is just the tip of the iceberg where food is wasted. As a teenager, we are eager to apply the knowledge we have learned in school in science and technology to turn the tip of the iceberg into a flat bottom and reduce the occurrence of one of the factors that cause environmental damage.

This project is also a successful completion and over-fulfilled within the planned time, which is the greatest gift to the producer. In this meaningful project, we learned much new knowledge and realized the importance of teamwork. On our study trip, we were deeply attracted to the power of science and technology. We found that technology can make our life more convenient and help us solve social problems as much as possible. Although this technology cannot ultimately reduce the huge amount of food waste, through our trial, it is really to form the double convenience of our life. We remind ourselves every day to protect the environment and reduce food waste.

Nearly one billion tons of food is wasted every year, about twice the best estimate in the past, according to the Food Waste Index Report, published by the United Nations Environment Programme (UNEP). On the fourth day of this month. This report is the most comprehensive to date, which points out that the world's household food waste amounts to 74 kg per person per year. Unfortunately, food waste has continuously been rising, and this scheme does not claim to reduce food waste significantly. Still, hopefully, it can be used to remind the public that it is necessary to start taking action to reduce food waste.

This scheme can also be optimized. For example, it can also be widely used in industrial activities, social security, elderly services, etc. It is believed that after the popularization of the intelligent refrigerator assistant in the future, the food waste caused by improper storage methods can be significantly reduced, food safety can be guaranteed for the user, and the goal of improving life intelligence can be achieved.

Food waste has become a significant problem in the development of the world. We believe the proposal can greatly reduce this situation, and the world will improve with the cooperation of human beings and machines.

Bibliography

Faster R-CNN—Torchvision 0.15 documentation. (n.d.). https://pytorch.org/vision/stable/models/faster_rcnn.html

Food Waste Index Report 2021. (n.d.). https://wedocs.unep.org/bitstream/handle/20.500.11822/35280/FoodWaste.pdf

Global Food Waste. (n.d.). https://ls.chiculture.org.hk/tc/reform-info/583

Gould, L. G., Skye. (n.d.). The world waste the most food image source. *Business insider*. https://www.businessinsider.com/which-parts-of-the-world-waste-the-most-food-2016-3

Ren, S., He, K., Girshick, R., & Sun, J. (2017). Faster R-CNN: Towards real-time object detection with region proposal networks. *IEEE Transactions on Pattern Analysis and Machine Intelligence, 39*(6), 1137–1149. https://doi.org/10.1109/TPAMI.2016.2577031

Kin Un Lam is teacher at Pui Ching Middle School, Macau. He has been working in software engineering for more than ten years. He enjoys the value and significance of making progress together with his classmates.

Chapter 6
A Rotor Fault Detection System Based on Nonlinear and Dynamic Response

Jiawei Huang, Baoqing Huang, Jiayu Huang, and Yixing Liang

Abstract This research proposes a detection system that utilizes electromagnetic induction to accurately diagnose rotor faults in motor and quality while minimizing noise interference in production and maintenance environments. The system is designed to improve the efficiency of factory inspection lines and prevent catastrophic motor failure accidents, thereby ensuring personal safety. The system employs a time-varying magnetic field to excite the rotor under test, which generates an electromagnetic response associated with rotor failure or quality issues. The response is then converted into an acoustic signal processing for fault detection. This approach offers a comprehensive solution for accurately detecting rotor faults and enhancing the overall quality of motor production . Additionally, the system can be utilized in various industrial applications, including machinery manufacturing, power generation, and transportation. Overall, the proposed system offers a reliable and efficient solution for fault detection, which is crucial for ensuring safe and reliable motor operation in various industrial settings.

Keywords Rotor fault diagnosis · Ensemble Empirical Mode Decomposition (EEMD) · Noise reduction

Highlights

This research proposes a comprehensive solution for accurately detecting rotor faults and enhancing the overall quality of motor production. By utilizing electromagnetic

---

Lou Hau High School, Macao was founded by a labor union in the 1950s. The school adheres to the teaching reform for a long time and seeks progress in the reform. In recent years, the Teaching Reform Committee has been set up to carry out diversified intelligent evaluation, and constantly study and analyze the theoretical viewpoints, concepts and measures of education at home and abroad, so as to achieve the results of creative thinking teaching and quality education.

J. Huang · B. Huang · J. Huang · Y. Liang (✉)
Lou Hau High School, Macao, China
e-mail: liangyixing@yahoo.com.hk

J. Huang
e-mail: huangjia22@mails.tsinghua.edu.cn

R. Huang et al. (eds.), *A Collection of AI Innovations by Chinese Teenagers*, Smart Computing and Intelligence, https://doi.org/10.1007/978-981-97-3316-3_6

induction and acoustic signal processing, this system offers a reliable and efficient solution for fault detection, which is crucial for ensuring safe and reliable motor operation in various industrial settings.

Challenges The aim is to enhance the operation of factory inspection lines to prevent catastrophic motor failures, thereby ensuring user safety and improving efficiency.

Solution A comprehensive inspection system was developed based on electromagnetic induction, which is capable of accurately diagnosing rotor faults and quality in motors/generators in both production and maintenance environments.

Outcome The system could diagnose motor rotor faults with near 100% detection rate after workmanship modifications. The FM1388 core and the EEMD algorithm aided in analysis, effectively reducing noise or interferences, and providing a high-fidelity extracted vibration signal.

How Does the Proposed System Detect Rotor Faults in Motors?

The system works by utilizing a magnetic testing platform to observe the magnetic physical vibration phenomenon of the rotor. The rotor is excited by a time-varying magnetic field, which generates an electromagnetic response associated with rotor failure or quality issues. The response is then converted into an acoustic signal processing for fault detection. The acoustic signals of rotor vibration are sampled and processed by a DSP chip, and the background noises are removed. The original vibration signals from the electromagnetic probe are extracted. By analyzing the amplitude of spectrum data, the degree of the faultiness of the inter-turn short circuit of the rotor is further derived by referencing. The system employs a comprehensive detection system based on electromagnetic induction to minimize noise interference in production and maintenance environments, which ensures accurate fault diagnosis and analysis.

How Does the System Minimize Noise Interference in Production and Maintenance Environments?

The system utilizes a DSP chip to sample and process the acoustic signals of rotor vibration, which effectively eliminates high-frequency noise interference while retaining more fault details. Additionally, the system employs a comprehensive detection system based on electromagnetic induction to minimize noise interference in production and maintenance environments, which ensures accurate fault diagnosis and analysis.

What Are the Potential Industrial Applications for this System?

The system can be utilized in various industrial applications, including machinery manufacturing, power generation, and transportation. It can be used for Motor Production Inspection (Large-Sized Rotors for Power Generation), electric vehicles, and power generation. The system offers a reliable and efficient solution for fault detection, which is crucial for ensuring safe and reliable motor operation in various industrial settings.

6.1 Project Overview for Detecting Rotor Faults in Motor

The research you mentioned is focused on developing a comprehensive detection system for electric motors and generators, with a specific emphasis on diagnosing faults in the rotor windings. Electric motors and generators are critical components in various applications, and ensuring their safe and efficient operation is essential for both the machines themselves and the safety of users. With the increasing use of electric vehicles and wind power generation, the need for accurate fault diagnosis and maintenance has become more significant.

Here is a breakdown of the key components and goals of this research:

(1) **Rotor Winding**: The rotor winding is identified as a crucial component of electric motors and generators. Faults in the rotor winding, particularly short-circuit faults, can have a significant impact on the performance and safety of these machines. Therefore, detecting and diagnosing rotor winding faults is a primary objective of this research.

(2) **Detection System**: The research aims to develop a comprehensive detection system that utilizes electromagnetic induction principles. This system is likely designed for use in both production and maintenance environments, suggesting its potential for diagnosing faults during manufacturing and ensuring ongoing reliability during the operational lifespan of these machines.

(3) **Signal Analysis**: The research incorporates signal processing techniques to analyze the vibrations of the rotor. Specifically, it mentions the use of the EEMD (Ensemble Empirical Mode Decomposition) algorithm for noise reduction. This technique decomposes the original vibration signals into simpler components, potentially making it easier to identify and diagnose faults accurately.

(4) **Noise Reduction Chip**: A noise reduction chip is proposed as part of the signal processing approach. This chip likely plays a crucial role in improving the quality of the signal data collected from the rotor, making it easier to distinguish between normal operation and fault conditions.

(5) **Signal Segmentation**: The research mentions the goal of reducing signal segmentation to the maximum extent. This implies a desire to streamline the analysis process, making it more efficient and effective in diagnosing rotor winding faults.

(6) **Wavelet Transformation**: The research acknowledges the limitations of wavelet transformation in this context. It is likely that the proposed EEMD-based approach is intended to overcome these limitations and provide more accurate fault diagnosis.

In summary, this research project aims to develop an advanced detection system that utilizes electromagnetic induction and signal processing techniques to accurately diagnose faults in the rotor winding of electric motors and generators. This technology has the potential to improve the safety, reliability, and performance of these machines, which are essential in various industries, including electric vehicles and wind power generation.

6.2 Project Design of Detecting Rotor Faults System

6.2.1 Structure of This System

This project consists of as following part: motor rotor (type, size, power, and the number of turns are not restricted); electromagnetic probe (thin sheets of ferro magnetic material); electromagnetic generator (230 V AC, to excite the inter-turn of the rotor windings); test control pillar stand (to hold the stationary controller in place); stationary controller (a hold that fits the electromagnetic probe); miniature stepper motor (for controlling electromagnetic probe to complete vertical detection); dual controller (to ensure the working stability of the stepper motor and reduce the delay of remote control launching); wireless remote control (a transmitting antenna can be fitted, for stepping motor to control the vertical operation of electromagnetic probe); sound level sensor (Bluetooth connection is available, function as data collection); virtual oscilloscope (i-Seekwhy, convenient data processing and waveform transmission); electrical control panel (which housed the test control pillar stand's wireless remote controller and stepper motor driver to controls the electromagnetic probe's movement to provide up, down, and stop movement); computer (Served as terminal output to process and display i-Seekwhy and sound level sensor); Noise reduction chip (Speech IC size: 7 × 7 mm, demo board, recording board. Model: FM1388. Number of package outlines: TQFP-48pin. Integrated ForteMedia technology 2 stereo digital microphone ports).

6.2.1.1 Noise Reduction Chip

Chip Structure

Its package shape is TQFP (thin square flat package) (Figs. 6.1, 6.2 and 6.3).

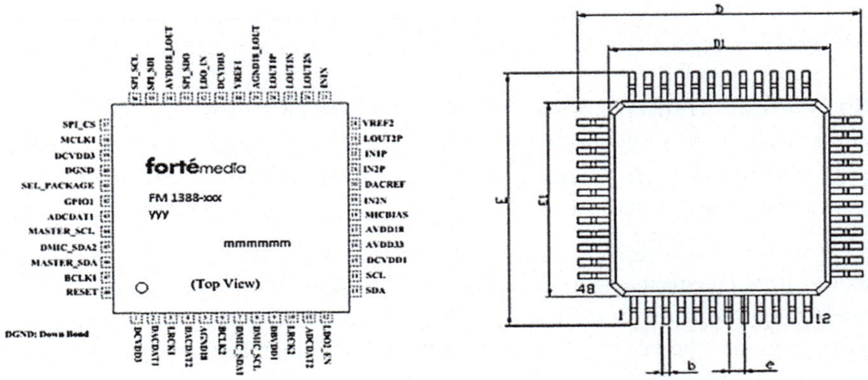

Fig. 6.1 TQFP-48pin configuration

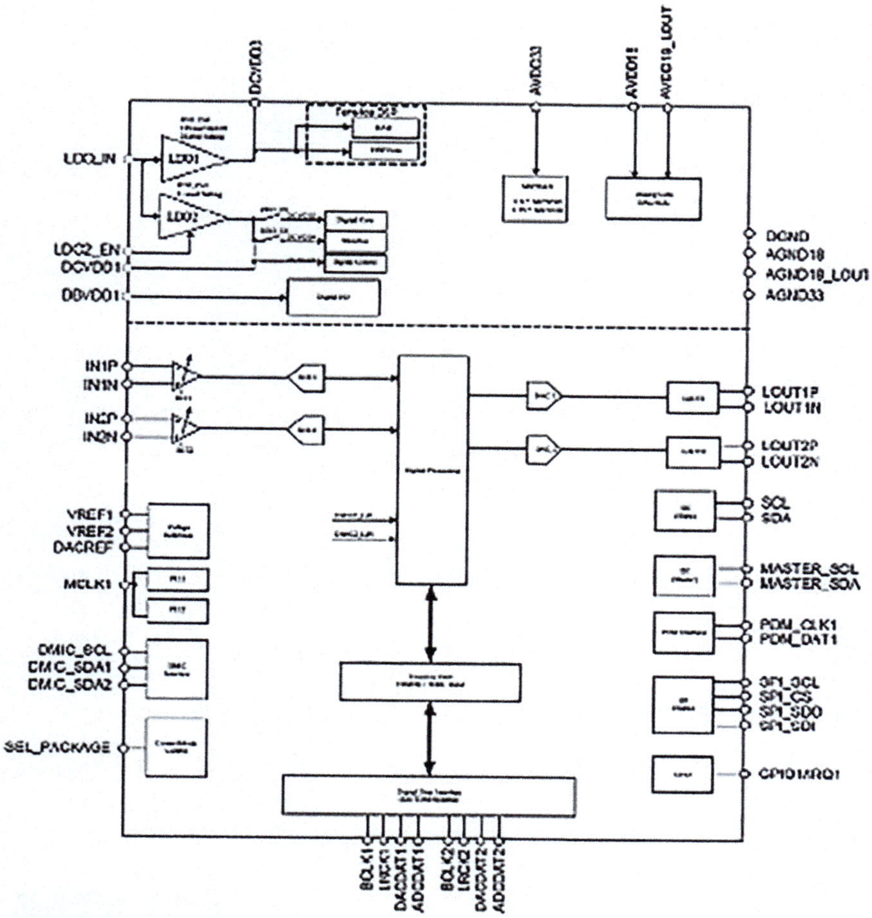

Fig. 6.2 IC hardware block diagram

Chip Function

The noise reduction chip integrates the latest Fortemedia technology to eliminate ambient noise and acoustic echo and keeps the rotor's original vibration signal natural in various noisy environments to improve signal clarity (Figs. 6.4, 6.5 and 6.6).

6.2.2 Research Methods of This System

The research experiment is divided into two stages: Stage 1—Automated Quality Inspection System for Rotor Coils of Commutator Motor; Stage 2—Motor Rotor Fault Detection and its Signal Processing & Analysis Using EEMD.

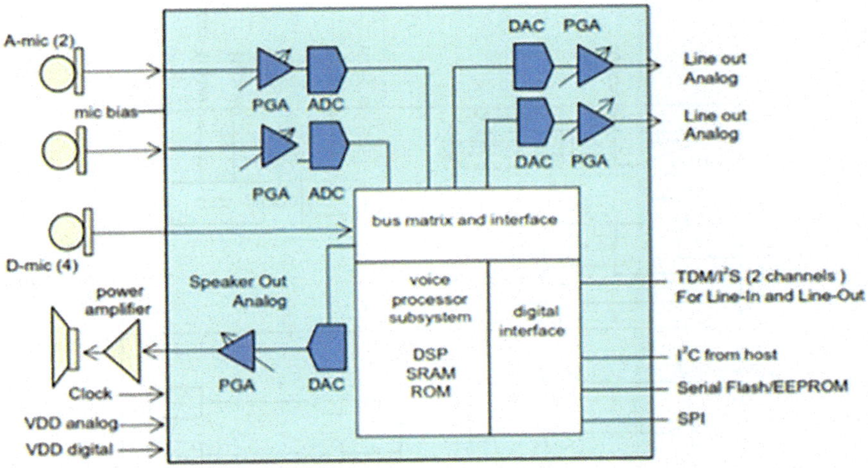

Fig. 6.3 Chip structure diagram

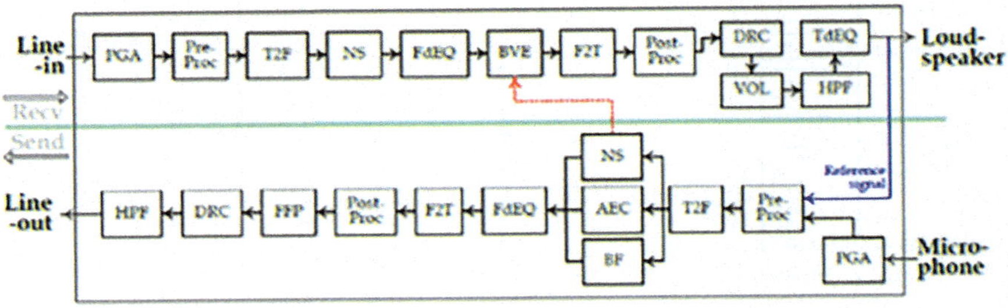

Fig. 6.4 Signal processing flow

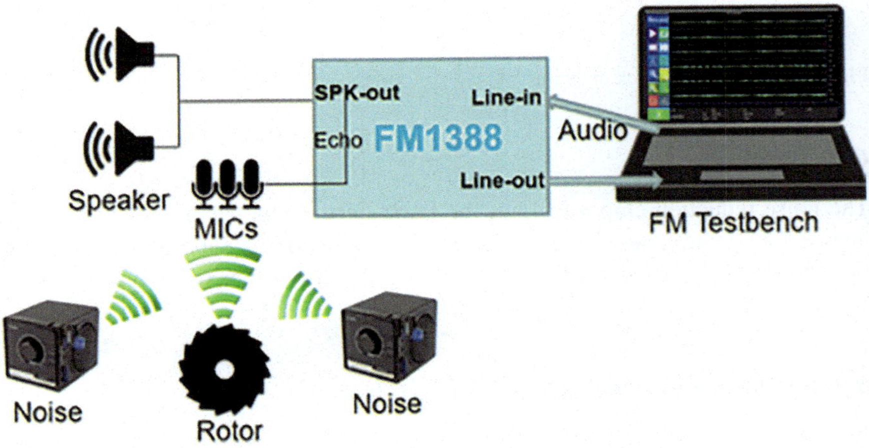

Fig. 6.5 Schematic diagram

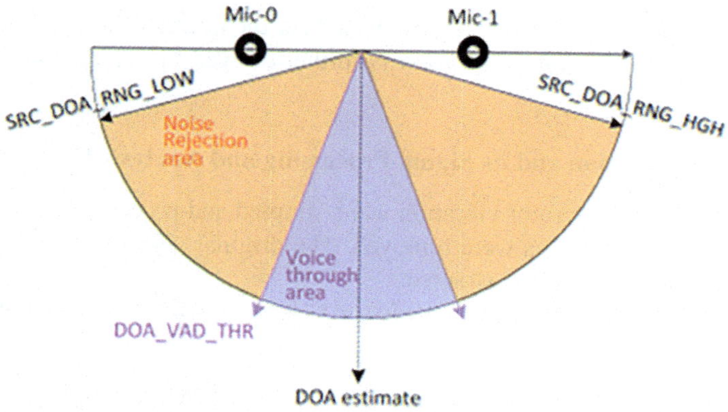

Fig. 6.6 The direction of arrival

Automated Quality Inspection System for Rotor Coils of Commutator Motor

By establishing a magnetic testing platform to observe the magnetic physical vibration phenomenon of the rotor, and further, with the help of sound sensors and virtual oscilloscopes, rotor fault signals can be collected and analyzed from the sound vibration spectrum.

The experiment steps are as follows:

1. Assemble the electromagnetic generator and probe on the test control pillar stand with the manual probe control mechanism.
2. Install a small stepping electric unit at the connection of the handle of the test control pillar stand. A remote lifting control for the probe is used in place of the manual lifting handle. This improved the robustness of the device and allowed for more consistent measurement.
3. Optimize the placement of the testing rotor between the grooves of the electromagnetic generator.
4. Employ the use of remote control to decide the moving of the electromagnetic probe.
5. We are utilizing a sound level sensor and oscilloscope i-Seekwhy to collect the rotor vibration spectrum data reflected by the electromagnetic probe.

   7 Steps of using the sound level sensor

   (a) Operate the control panel to start the detection system.
   (b) Select the signal generator at the signal source.
   (c) The signal generator has two sound channels. Select the sine wave.
   (d) Open two soundtracks in the oscilloscope window.
   (e) Display the two-sound track waveform and the synthesized waveform.
   (f) Display the real-time frequency of the signal in the FFT window and frequency meter.
   (g) Store data in the control panel.

6. By analyzing the amplitude of spectrum data, the degree of the faultiness of the inter-turn short circuit of the rotor is further derived by referencing (Fireteanu et al., 2014).

Rotor Fault Detection and its Signal Processing and Analysis Using EEMD

The acoustic signals of rotor vibration were sampled and processed by a DSP chip, and the background noises were removed. The original vibration signals from the electromagnetic probe were extracted.

For the method used in stage 2, the first six steps are the same as in stage 1. Steps 1 and 2 are as follows:

1. We are using the noise reduction chip to analyze the azimuth inductance waveform of the winding frequency spectrum of the electric motor.
2. We are collecting the rotor's fault vibration signals and extracting the rotor's original vibration signals. First, the noise reduction chip filters the noise in the signal. Then, the spectrum amplitude of the filtered rotor fault signal is analyzed to detect whether the rotor is short-circuited and the fault degree of the original signal. Furthermore, the algorithm identifies the quality of the rotor (Lei et al., 2009).

6.3 Experimental Results

6.3.1 Automated Quality Inspection System for Rotor Coils of Commutator Motor

6.3.1.1 Intuitive Detection Methods

Through the experiment, the current quality of the rotor can be easily judged by the method of hierarchical rating, but the accuracy is not enough. They are divided into 10 experimental groups. A total of 10 rotors and 120 slots were tested. Test results (rating and quantity) are A (32), B (30), C (17), D (29), E (6), and F (6) (Tables 6.1 and 6.2).

Advantage

To ensure the detection accuracy of the initial stage. Using intuitive detection methods (Visual, Auditory, Tactile) and discriminant methods (Applicable to Stage 2 below). Then, referring to the research conclusions of Wan et al. (2003) and judging from the experimental results, the definition of rotor fault degree is formulated.

Inadequacy

Using intuitive detection methods (visual, auditory, tactile), standards can also be set to judge rotor quality artificially. The detection degree is determined according to the

Table 6.1 A brief description of motor rotors' quality manual approval evaluation

Amplitude intensity of electromagnetic probe	Rotor quality assessment	Need to do
Strong	Breakdown	It should be replaced and repaired immediately
Under Strong	Short.Circuit (Seriously)	Should be further repaired
Average	General	Should self-repair the rotor core winding short circuit, saving money
Upper Slight	Short.Circuit (Slightly)	One or the twelve turns of iron core coil winding may have a short circuit, should self-repair, saving money
Slight	Reliable	Temporally stable
No	Off Size	No need to worry about the short circuit of the rotor. It is stable

Table 6.2 General data of a hundred experiments

Group	Groove											
	1st	2nd	3rd	4th	5th	6th	7th	8th	9th	10th	11th	12th
I	A	D	B	C	D	A	B	A	B	A	D	B
II	A	A	B	A	B	C	A	B	A	F	A	C
III	B	B	A	A	C	A	B	B	D	C	D	A
IV	D	A	B	A	B	A	D	B	A	B	D	E
V	A	B	D	A	C	D	D	D	D	A	A	A
VI	C	D	D	C	D	A	B	A	F	C	B	D
VII	B	A	B	E	A	D	B	D	B	8	D	C
VIII	D	E	A	D	C	D	D	C	B	D	F	C
IX	D	A	F:	A	B	B	C	D	F	E	C	B
X	A	8	E	F	C	E	D	B	A	D	C	B

The amplitude of the electromagnetic probe (A: Strong amplitude B: Under strong amplitude. C: Average amplitude D: Upper slight amplitude E: Slight amplitude F: No amplitude) (I, II, III, IV, V, VI, VII, VIII, IX, X—Each experimental group) (1, 2, 3, 4, 5, 6, 7, 8, 9, 10, 11, 12—Slot of each rotor)

electromagnetic probe's vibration amplitude and frequency size, but the accuracy is insufficient.

Due to environmental noise, sound volume is difficult to hear in a noisy environment, so the degree of rotor fault cannot be judged accurately. The electromagnetic probe's contact area with the rotor core, angle, and time limit has the most significant influence on the accuracy of the experiment.

6.3.2 Automatic Method

The experiment was divided into 10 experimental groups. A total of 10 rotors with 120 slots were detected, of which 43 slots were short-circuited. The probability of a short circuit is 35.8%. The results were classified into short-circuit and non-short-circuit groups. The short-circuit group and the non-short-circuit group were also divided into two groups (Figs. 6.7, 6.8, 6.9 and 6.10).

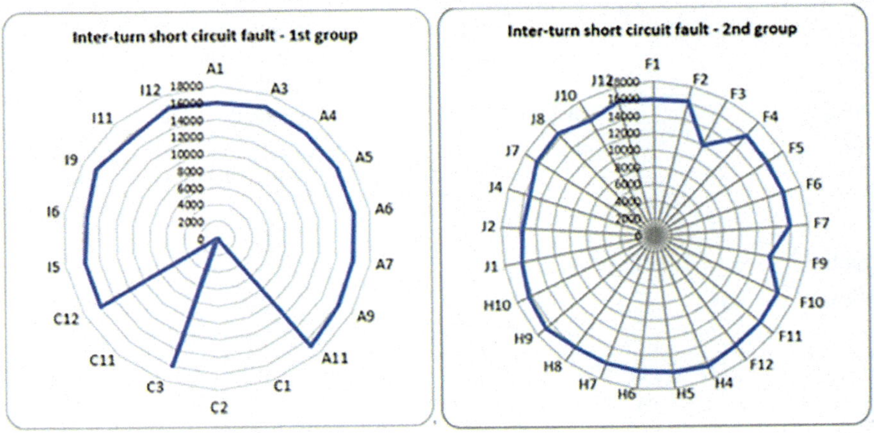

Fig. 6.7 The 1st and the 2nd group inter-turn short-circuit fault inductance spectrum analysis

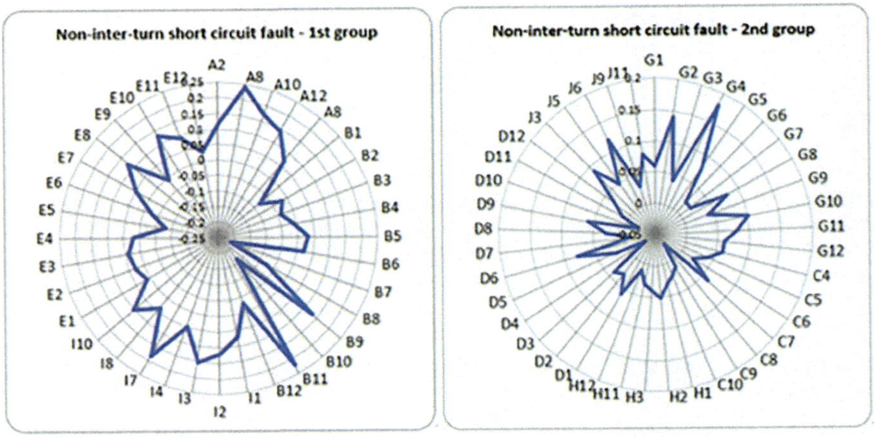

Fig. 6.8 The 1st and 2nd group non-inter-turn short-circuit inductance spectrum analysis group

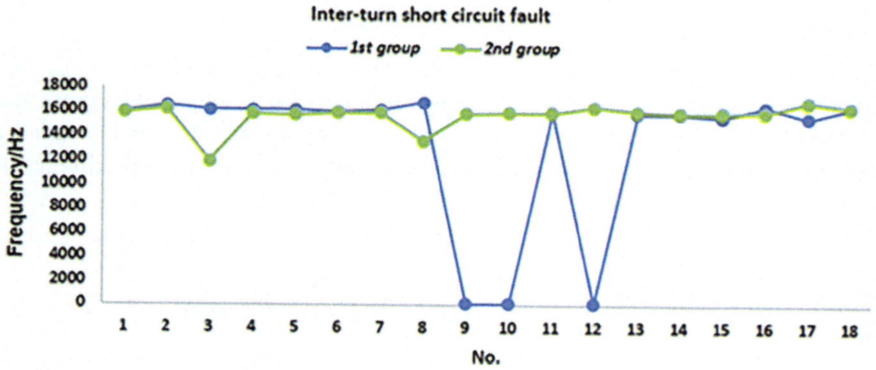

Fig. 6.9 Spectrum linear diagram of the motor rotor winding's inter-turn short-circuit fault group

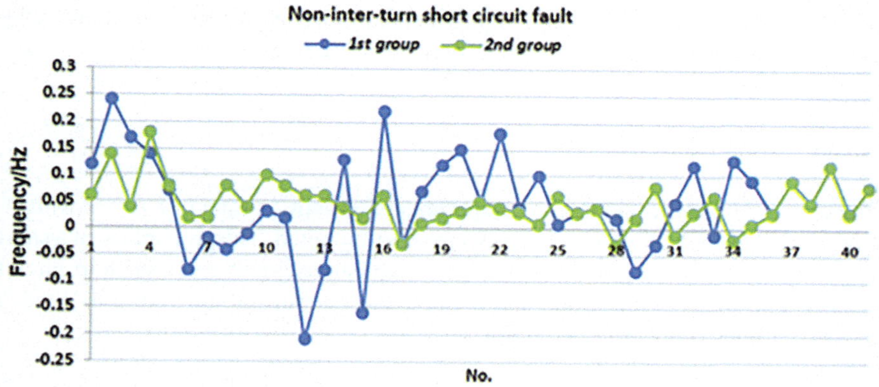

Fig. 6.10 Spectrum linear diagram of the motor rotor winding's non-inter-turn short-circuit fault group

Improvement

The automatic method no longer adopts the intuitive detection approach but uses the sound level sensor to collect the sound signal of the rotor vibration. And then judging the short-circuit condition of the rotor by analyzing the data, which has higher accuracy and degree of automation.

Using the remote control to operate the electromagnetic probe lets the electromagnetic probe's contact area with the rotor core, angle, and time limit, which no longer influences the accuracy of the experiment.

Advantage

To minimize instability as much as possible, decrease the degree of controlling the electromagnetic probe with manual control.

Inadequacy

Due to the environmental noise when the sound level sensor collects the sound signal of the rotor vibration, it is easily affected by the environmental noise. As a result, the collected vibration signal of the rotor is not entirely pure, and it cannot accurately judge whether the rotor is short-circuited through the data, and the detection result is prone to considerable error.

6.3.3 Signal Processing and Analysis Using EEMD

A sound level sensor and oscilloscope i-Seekwhy have been used to improve the above deficiencies and collect the rotor vibration spectrum. It no longer adopts the intuitive detection method but uses the sound level sensor to collect the sound signal of the rotor vibration. And then judging the short-circuit condition of the rotor by analyzing the data, which has higher accuracy and degree of automation.

6.3.3.1 Data Analysis

The experiment was divided into 8 experimental groups. A total of 8 rotors with 96 slots were detected, of which 31 slots were short-circuited. The probability of a short circuit is 32.3%. The results were classified into short-circuit and non-short-circuit groups. The short-circuit group and the non-short-circuit group were also divided into two groups (Figs. 6.11 and 6.12).

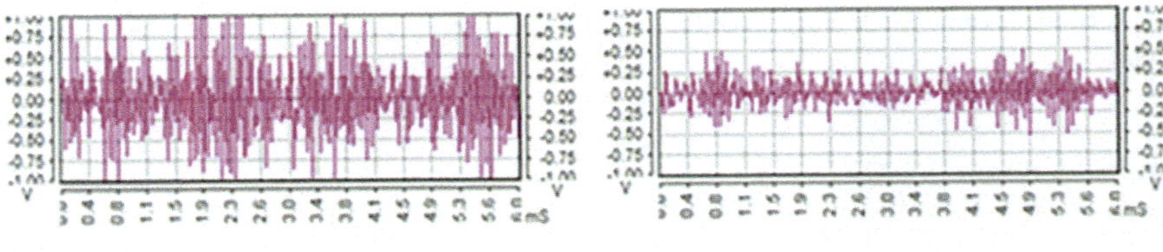

Fig. 6.11 Analysis of azimuth inductance waveform of the winding frequency spectrum of the motor rotor

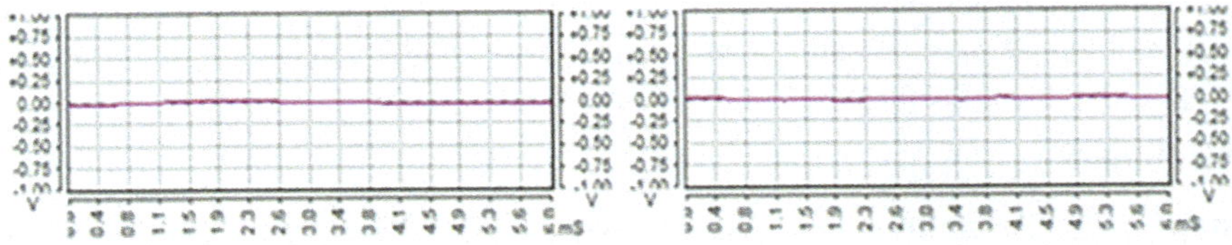

Fig. 6.12 Analysis of azimuth inductance waveform of the winding frequency spectrum of the motor rotor

Improvement

After chip processing, high-frequency noise interference to the rotor fault signal is effectively eliminated while retaining more fault details, conducive to subsequent fault diagnosis and analysis (Miao et al., 2020).

6.3.3.2 Advantage

Effects from environmental noise and noise interference have significantly been reduced. The acoustic signals of rotor vibration were sampled and processed by a DSP chip, and the background noises were removed.

The original vibration signals reflected on the electromagnetic probe were extracted. The experimental data and analysis results are more accurate and dramatically improve the rotor detection efficiency and accuracy.

6.4 Conclusion and Future Plans

Through experimental design, continuous project evaluation, project data analysis, innovative iterative research summarizing engineering methods, and the analysis of the rotor short-circuit vibration signal amplitude, this project has achieved its primary goal of accurately determining the rotor fault severity. The research started with the detection of alternating electromagnetic fields. The experiment was repeated to refine the detection process, and the results obtained reached the expectation in the end. The collected data are conducive to further research.

The experimental method proposed in this project can effectively improve the IMF (Intrinsic Mode Function) variable component caused by the suppression of wavelet transform on signal segmentation. Based on the principle of the EEMD algorithm, a noise reduction chip is proposed to decompose the original vibration signals of the rotor. It is not limited by time and can reduce the delay limit of signal segmentation by wavelet transform to the maximum extent, and then the original vibration signal of the rotors can be obtained. By removing the delay, the geometric characteristics of the original vibration signal can be well preserved. It is more effective than the general wavelet domain filtering method.

In this study, a comprehensive detection system based on electromagnetic induction is developed to minimize noise interference in production and maintenance environments and thus accurately diagnose motor rotor faults and quality. This research can improve the operation of factory inspection lines, prevent catastrophic motor failure accidents, and ensure the personal safety of users, which has great significance and value.

The automatic rotor fault inspection device developed in this project can be used for Motor Production Inspection (Large-Sized Rotors for Power Generation). Large rotors, such as those used in electric vehicles and power generation, are expensive, and their failures can cause massive safety and cost issues. In addition, one of the world's largest numbers of moto/generator rotor waste problems results in heavy metal pollution. This research provides a valuable and satisfactory response to the problem. Moreover, it is used to improve the production efficiency of the factory assembly line. Electric power tools have countless manufacturing, construction, and facility maintenance applications. Also, it can be used to repair motor loss and prevent mechanical metal contamination.

The future engineering goal of this project will focus on researching and developing a set of comprehensive quality inspection systems for electric vehicle starting commutators. It is known that this can be done from the current extension principle. By extracting and analyzing the original vibration signal of the rotor, we aim to maximize the application of functional modules of the noise reduction chip. Optimize the storage layer of signal data and improve the standard connection for automatically managing changing data structures.

References

Fireteanu, V., Lombard, P., & Constantin, A. I. (2014). Detection of a short-circuit fault in the stator winding of induction motors through neighboring magnetic field harmonics. In *2014 International conference on electrical machines (ICEM)* (pp. 1555–1561). https://doi.org/10.1109/ICELMACH.2014.6960389

Gaeid, K., & Ping, H. (2011). Wavelet fault diagnosis and tolerant of induction motor: A review. *International Journal of Physical Sciences*. https://www.semanticscholar.org/paper/Wavelet-fault-diagnosis-and-tolerant-of-induction-A-Gaeid-Ping/e113ff9aa63656b504a45fbc3301a44b6f269de1

Lei, Y., He, Z., & Zi, Y. (2009). Application of the EEMD method to rotor fault diagnosis of rotating machinery. *Mechanical Systems and Signal Processing, 23*(4), 1327–1338. https://doi.org/10.1016/j.ymssp.2008.11.005

Miao, F., Zhao, R., & Wang, X. (2020). A new method of denoising of vibration signal and its application. *Shock and Vibration, 2020*, e7587840. https://doi.org/10.1155/2020/7587840

Wan, S., Xu, Z., Li, Y., Hou, Z., & Li, H. (2003). Analysis of generator vibration characteristic on rotor winding interturn short circuit fault. In *Sixth International conference on electrical machines and systems, 2003 (ICEMS 2003)* (Vol. 2, pp. 882–885).

Yixing Liang Vice teaching director of The Macau Workers' Children High School since 2003 and he also serves as a member of the Secondary School Committee of the Chinese Physical Society, president of the Macau Physics and Education Research Society, president of the Macau Technicians Association, president of the Macau Youth Sailing Model Association, and vice president of the Greater Bay Area Innovation and Invention Association.

Chapter 7
A Detection System for Choosing Ripe Strawberry with Machine Learning

Yuxia Lin, Siming Huang, Yi Zheng, Xu Zheng, Zijun Huang, and Zhihao Zhuo

Abstract The timely and accurate detection of fruit ripeness is critical in preserving food safety and maintaining market order. The project focuses on enhancing the detection and accuracy rate of strawberries: a popular fruit with a short shelf life and is highly perishable. The project employs machine learning algorithms and computer vision analysis to prompt recognition of the ripeness stages of strawberries. The project analyzes images of strawberries and uses mathematical models to categorize the fruit into their respective ripeness stages. Through the use of machine learning, the project aims to improve the efficiency and accuracy of fruit classification, resulting in better quality control and a reduction in food waste. Furthermore, by utilizing computer vision analysis and machine learning algorithms, the project can aid in the timely harvesting, processing, and distribution of strawberries, ultimately benefiting the agricultural industry and consumers alike. The project presents a promising solution for the future of the fruit industry and food safety.

Fuzhou No.3 High School is located on the shore of the scenic West Lake. Founded in 1942, the predecessor was "Fuzhou City Junior High School". In June 1952, after the unified adjustment of the school name in Fuzhou, it was officially named "Fuzhou No. 3 High School".

Y. Lin · S. Huang (✉) · Y. Zheng · X. Zheng · Z. Huang · Z. Zhuo
Jin'an Campus of Fuzhou No. 3 High School, 350000 Fujian, China
e-mail: hsm_fz@qq.com

Y. Lin
e-mail: 1542986837@qq.com

Y. Zheng
e-mail: 1226651518@qq.com

X. Zheng
e-mail: 2069004950@qq.com

Z. Huang
e-mail: 1542986837@qq.com

Z. Zhuo
e-mail: 1542986837@qq.com

R. Huang et al. (eds.), *A Collection of AI Innovations by Chinese Teenagers*, Smart Computing and Intelligence, https://doi.org/10.1007/978-981-97-3316-3_7

Keywords Strawberry ripeness · Machine learning · Support vector machine (SVM) model · Rapid detection method

Highlights *This project aims to improve the efficiency and accuracy of fruit classification, resulting in better quality control and a reduction in food waste. By utilizing computer vision analysis and machine learning algorithms, this project can aid in the timely harvesting, processing, and distribution of strawberries, ultimately benefiting the agricultural industry and consumers alike.*

Challenges: The food industry has been challenged by "low detection efficiency" and "food adulteration" in recent years. This study uses strawberries to focus on improving detection speed and accuracy through computer vision and machine learning, facilitating fruit farmers during harvest and broadening machine learning applications in the food sector.

Solution: Experienced fruit farmers sorted harvested strawberries into four ripeness categories, with 150 images captured for each. We utilized support vector machine (SVM) for classification and compared the results with those from a convolutional neural network (CNN).

Outcome: A set of strawberry imaging equipment was developed. An advanced strawberry ripeness detection model was created.

What machine learning algorithms are used in this project?

This project utilizes two machine learning algorithms: traditional machine learning and deep learning. The traditional machine learning model used is SVM, which achieved an accuracy of 97.50%. The deep learning model used the method of transfer learning of convolutional neural network (CNN), and the classification accuracy of the final model is 94.28%.

How does this project benefit the agricultural industry and consumers?

This project benefits the agricultural industry and consumers by improving the efficiency and accuracy of fruit classification, resulting in better quality control and a reduction in food waste. By utilizing computer vision analysis and machine learning algorithms, the project aids in the timely harvesting, processing, and distribution of strawberries. This ultimately benefits the agricultural industry by increasing productivity and reducing costs, while also benefiting consumers by providing them with fresher and safer fruits.

Can this detection system be applied to other types of fruits?

While this project focuses on enhancing the detection and accuracy rate of strawberries, the methodology and techniques used in this project can be applied to other types of fruits as well. The project employs machine learning algorithms and computer vision analysis to prompt recognition of the ripeness stages of strawberries. The project analyzes images of strawberries and uses mathematical models to categorize the fruit into their respective ripeness stages. Therefore, with appropriate modifications, this detection system can be applied to other types of fruits as well.

7.1 Overview of Choosing Ripe Strawberry Project

In recent years, low detection efficiency and food adulteration have become problems that restrict the development of the food industry. In this thesis, the strawberry is taken as the research object, and the problems of improving the detection and accuracy rate of the strawberry need to be focused on and solved. To provide a rapid detection method to protect food safety and maintain the market order, this thesis adopts the computer vision analysis method based on machine learning to carry out the relevant research on strawberry maturity recognition. It expands the application of machine learning in the field of food.

Based on the research on strawberry maturity, it may bring great convenience to fruit growers during the picking period to avoid the corrosion of late picking in the transportation process or the awful taste brought to customers by early picking. We built a simple strawberry collection device to collect four different maturity data sets, which divide the training set and the test set in a ratio of 8: 2. Then, the traditional machine learning model of support vector machine (SVM) is established, and the accuracy is 97.50%. The deep learning model is built in the second part, and the model's classification accuracy is 94.28%. In this way, the strawberries are picked in stages according to the actual market requirements, the corrosion loss of the strawberries can be reduced, the cost is reduced, and the production efficiency is improved.

With the development of the digital economy, the big data era is entering people's lives in an unstoppable manner and gradually permeating various industries and fields. As an advanced big data analysis technology, machine learning has many successful cases in image processing, spectral analysis, speech recognition, and target detection. For example, transport companies use machine learning to support package processing, and cranberry manufacturers use machine learning to enhance operations.

At the same time, people's demand for food quality is increasing. Therefore, the project is committed to combining food detection with machine learning and uses big data analysis technology to solve the current problems and challenges in the food field.

Strawberries are easy to spoil and not easy to store because of their tender flesh. In general, once strawberries need to be transported or stored over long distances, fruit growers pick them when they are about to ripen, while locally sold strawberries choose to pick them when they are fully ripe. It avoids losses from strawberry corrosion during transportation and reduces food safety risks. Therefore, the differentiation of strawberry maturity degree is an important task of the fruit industry. The fruit maturity degree index informs harvest logistics and supply chain management decisions to provide consumers with the best fruit quality. Indexes used to evaluate fruit maturity degree include soluble solids concentration, flesh firmness, starch concentration, titratable acidity, skin and flesh color, fruit size and shape, ethylene production, and respiration rate. Due to the particularity of strawberry fruit, it is customary to use the ratio of the red area on the surface of strawberry flesh as an index to distinguish strawberry maturity. According to the industry standard, the ripening of strawberry fruit can be divided into four stages according to the coloring area: green ripe stage (25%), white ripe stage (50%), color transformation stage (75%), and red ripe stage (full coloring).

Many researchers have reported on their work in predicting the ripeness of different fruits. Tao Huawei et al. first used the color complete local binary patterns to extract the image texture features and then used the color histogram to extract the image color features. The matching score fusion algorithm was used to fuse the color and texture features, and then the fruit types were recognized. Finally, the experimental results show that compared with other feature extraction methods, this method can reduce the time overhead and improve the recognition rate. Zawbaa et al. used the shape and color of fruits to identify each image feature. The system was divided into three stages: preprocessing stage, the feature extraction stage, and the classification stage. The preprocessing stage reduces the color index. The feature extraction stage generates a feature vector for each image in the data set using scale-invariant feature transform (SIFT) and shape and color features.

Moreover, the k-nearest neighbor classification and support vector machine (SVM) algorithm are used to identify fruit species in the classification stage. The experimental results show that the method can accurately identify 178 different kinds of fruits. Most fruit classification and recognition algorithms often use images in a strictly limited environment, eliminating the influence of the external environment on the pictures. However, the actual environment is complicated, as the color and texture characteristics of the fruit image become complicated along with the different growth cycles of the fruits. Furthermore, the natural light intensity variations and the shadow are different when the fruit images are acquired, increasing the difficulty of fruit classification and recognition, so better classification and recognition methods are needed to solve these problems. Hossain et al. proposed an efficient framework for fruit recognition based on deep learning. Some research data show that the fine-tuned classical network model positively affects fruit and vegetable classification. Li et al. used the improved VGGM network model to classify fruits and vegetables. The experimental results show that the convergence speed and precision of the network are improved by combining VGGMBN with the batch processing normalization layer, and the classification accuracy of fruit and vegetable classification is 96.5% compared with the VGG network and AlexNet network. Compared with the traditional machine learning method, deep neural network reduces the process of manual intervention and possesses the characteristics of autonomous learning and expressive ability. Many scholars are also studying deep learning techniques in agriculture increasingly.

Traditionally, the artificial grading of strawberries based on human visual perception is time-consuming and expensive. Furthermore, the classification and quality assessment of fruits by visual inspection can lead to errors due to external influences such as fatigue and subjective bias. This paper proposed a simple, fast, and effective recognition strategy to classify strawberry maturity states using traditional machine learning and deep learning algorithms. This method provides a new direction for expanding the application of machine learning in the field of food.

7.2 Design and Implementation of Experiments

7.2.1 Experiment Design

The experiment is divided into four parts: First, strawberry image acquisition: first, place a portable LED light-supplementing studio on the table, and put an objective table in the middle of the studio to place strawberries. Fix the mobile phone bracket. Connect a computer outside the mobile phone. A fixed LED light strip is inside, right above the studio. The equipment is shown in the Fig. 7.1. The height of the fixed support is set to 28 cm, and the background of the studio is white, ensuring the stability shooting environment's stability and reducing the experiment's systematic errors. Then, a picture of the strawberry is taken by the mobile phone, transferred to the connected computer, and then the next step of processing is carried out.

Secondly, collecting the original images of strawberries: According to the industrial standard, the ripeness of strawberries can be divided into four stages: green ripe stage, white ripe stage, color transformation stage, and red ripe stage. Before identifying the ripeness, strawberries of different maturity degrees are picked, and experienced fruit growers are entrusted to help sort them into four types (i.e., green ripe stage, white ripe stage, color transformation stage, and red ripe stage) as the original data set category. Moreover, 150 strawberry images of each kind were taken to reduce the experimental error caused by the tiny color difference. Four categories have a total capacity of 600 images, which are saved in JPG format, and the image resolution is set to 224 * 224.

Thirdly, feature extraction is carried out on the strawberry image sample: based on the convolutional neural network (CNN) system, the convolution operation is carried out on the convolution kernel to realize the feature extraction of the strawberry image, which covers the characteristic information such as the color of the strawberry sample image.

Fourth, maturity matching: Using the model after machine learning, the strawberry images are matched with the 600 pictures in the library. Then, the computer

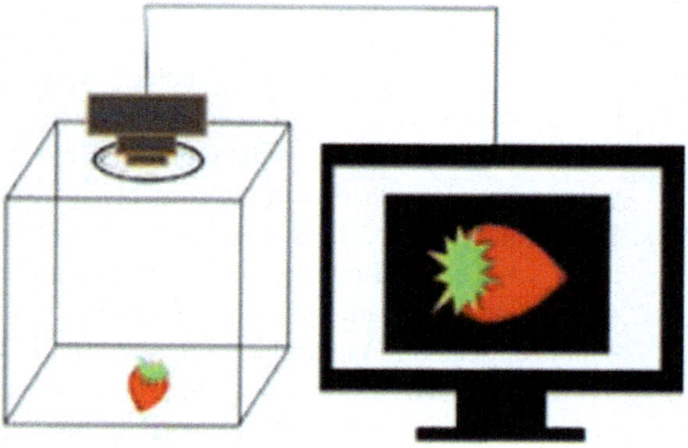

Fig. 7.1 Schematic diagram of image acquisition system

analyzes which maturity stage the strawberry images belong to. Moreover, the model parameters are modified, and the training model is trained in continuous judgment.

7.3 Processing of Strawberry Digital Images

7.3.1 Strawberry Image Processing Method

The color in the strawberry image is quantitatively represented using an RGB color model (color model refers to a mathematical model that uses a set of numerical values to describe the color quantitatively).

The first step is to segment the background in the original strawberry image.

(1) Obtain all original strawberry pictures (600 photos in step 2) under the RGB color model.
(2) The color histogram of each component image is obtained.
(3) Choose a reasonable threshold value for segmentation.

In that second step, the mean value and standard deviation of the R, G, and B components of the strawberry image in the RGB color model are extracted after the original image background is segmented.

In the third step, the mean values of R, G, and B components of 600 samples in the data set were visualized to have a more intuitive perception of the color characteristics of strawberries in different mature stages (Fig. 7.2).

7.3.2 Construction Steps of Machine Learning Model

The use of a training set and a test set is a common practice in machine learning and data modeling to assess the performance and validity of a model. Here's an explanation of the process you described:

(1) **Data set**: You mentioned that the data set consists of information related to strawberries. This data set likely includes various attributes or features of strawberries, such as size, weight, color, sweetness, and other relevant characteristics. Additionally, the data set might include a label or target variable indicating some property or classification of the strawberries, such as their ripeness or quality.
(2) **Hold-out method**: The data set is divided into two subsets: a training set and a test set. The hold-out method involves randomly splitting the data, typically with a predefined ratio. In your case, you used an 80:20 ratio, where 80% of the data is allocated to the training set, and 20% is allocated to the test set.
(3) **Training set**: The training set, which constitutes the majority of the data, is used to build or train your machine learning model. During this phase, the model learns from the patterns and relationships within the training data, aiming to

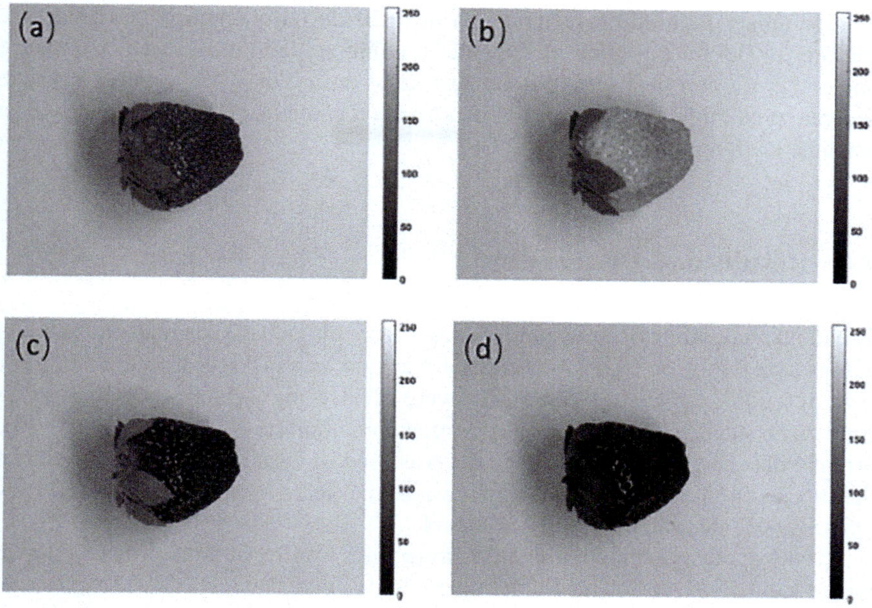

Fig. 7.2 Preprocessed image of strawberry. **a** Grayscale image, **b** R component image, **c** G component image, **d** B component image

capture the underlying characteristics that relate the input features to the target variable (e.g., strawberry quality).

(4) **Test set**: The test set is kept separate from the training data and is used to assess the model's performance. Once the model has been trained, it is applied to the test set, and its predictions are compared to the actual labels or values in the test set. This step helps evaluate how well the model generalizes to new, unseen data.

(5) **Model evaluation**: The model's correctness and error are assessed using various evaluation metrics, depending on the nature of the problem (e.g., regression, classification). Common metrics include mean squared error (MSE), accuracy, precision, recall, F1-score, and more. These metrics provide insights into how well the model performs on the test set.

(6) **Model validation**: By evaluating the model on the test set, you can validate its effectiveness and determine if it has learned meaningful patterns and relationships from the training data. If the model performs well on the test set and exhibits low error rates, it suggests that the model is a valid and reliable predictor.

(7) **Iterative process**: Depending on the evaluation results, you may need to iterate on the model, adjust its hyperparameters, or consider different algorithms to improve its performance. This process of model refinement and evaluation may be repeated until you achieve a satisfactory level of accuracy and reliability.

In summary, the hold-out method with an 80:20 ratio allows you to train a machine learning model on a portion of the data (training set) and assess its performance on another portion (test set). This process helps ensure that your model can make accurate predictions on new, unseen data and is a crucial step in developing robust machine learning models.

7.4 Results and Discussion

The rapid recognition of strawberry maturity recognition based on machine learning adopts two algorithm models: traditional machine learning and deep learning. The collected 600 samples of strawberries were divided into four maturity categories (ripe green stage, white ripe stage, color transformation stage, and ripe stage) as the original data set. The RGB color model was used to visually quantify the color of the images in the data set to construct a machine learning model. The strawberry maturity was identified by using the model.

All the applied technologies of the project are mature, and the project's digital image acquisition system device is simple, easy to build, and low-cost. The construction of the traditional machine learning model is very mature and easy to obtain and transform. The deep learning method can learn the advanced characteristics representing the input of the original data through a group of chained convolutional and pooling layers and extract the learned high-level features. The operation of the experiment process is quite simple. The rapid recognition accuracy rate of strawberry ripeness is up to 97.5%, which has the advantages of high efficiency and low cost. According to the project, fruit maturity can be identified more quickly and accurately than manual identification. The fruit can be picked in stages according to the market demand so that the cost is reduced and the production efficiency is improved. A new solution is also provided for the automatic grading of the fruits so that the loss caused by misjudgment during fruit picking is reduced, and people can taste more fresh fruits. The burden of fruit growers and the picking cost are reduced, which promotes the excellent development of the fruit industry.

Although the project uses machine learning and deep learning to quickly identify the ripeness of strawberries, due to the current technology, only a single strawberry can be identified each time, and large pieces of strawberries cannot be directly identified when used, which affects the identification efficiency. Therefore, there are still some technical difficulties to be solved in this project. In addition, to prevent the implementation of the project from being affected by site problems and limited by some objective conditions, we will continue to improve all aspects of the project to improve the accuracy and continuously break through the technical difficulties. In the subsequent experiments and studies, we will search for the problems in the actual implementation of the project and solve them.

After the project is put to a practical test, we will improve the number of samples according to the actual situation, find and solve more realistic problems, perfect the project, and make it with higher work efficiency, higher precision, and more relevant to the actual. We also hope that professionals will correct and help with this project.

The SVM model results are shown in Fig. 7.3. The horizontal coordinates are the sample number of the test set: 0 ~ 120. Thirty strawberry samples are randomly selected from each category as the test set. The vertical coordinates are the sample categories of the test set: 4 categories, 1 represents the white ripe stage, 2 represents the red ripe stage, 3 represents the green ripe stage, and 4 represents the color transformation stage, with a stair-stepping rise. One hundred twenty samples were randomly selected from each category to train the model, and the remaining thirty samples were used to evaluate the model's performance. From the figure, we can see that ripe white period: 2 strawberries are misjudged as color in the transformation stage; red ripe stage: 1 strawberry is misjudged in the color transformation stage; ripe green period: there is no misjudgment; in color transformation stage: there is no misjudgment.

As shown in Fig. 7.4, the confusion matrix diagram of deep learning is shown. The horizontal coordinates are the prediction label, and the vertical coordinates are the actual label. In each category, 594 pictures are randomly selected to evaluate the model. White ripe stage: two samples were misjudged to be the red ripe stage, ten samples were misjudged to be the ripe green stage, and twenty-nine samples were misjudged to be the color transformation stage; red ripe stage: Three samples were misjudged to be the white ripe stage, one sample was misjudged to be the ripe green stage, and fifty samples were misjudged to be the color transformation stage; green ripe stage: thirteen samples were misjudged to be the white ripe stage, and two samples were misjudged to be the red ripe stage; color transformation stage: sixteen samples were misjudged to be the white ripe stage, and ten samples were misjudged as the red ripe stage. The total classification accuracy of the test set is 94.28%.

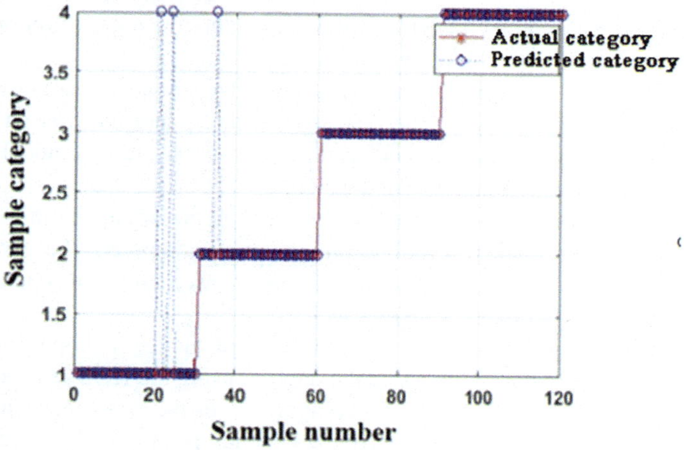

Fig. 7.3 SVM classifier

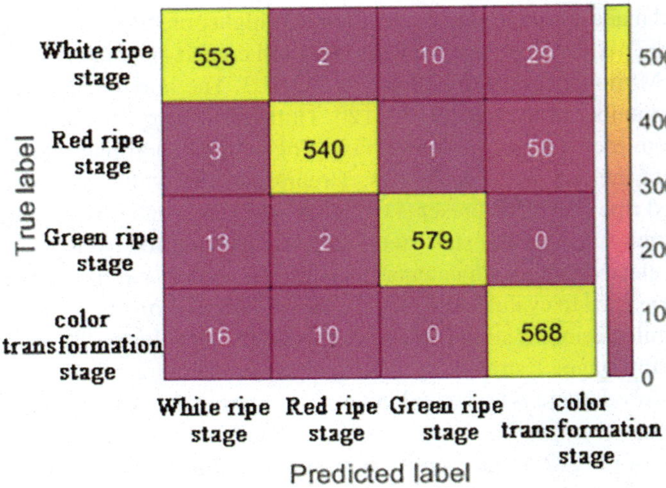

Fig. 7.4 Schematic diagram of deep learning confusion matrix

7.5 Conclusion and Future Plans

In this paper, a simple and rapid method of strawberry maturity recognition was proposed, i.e., a machine learning model was built to intelligently identify the ripeness of strawberries. First, a traditional machine learning model, SVM, is built with an accuracy of 97.50%. In the second part, a deep learning model is established, which adopts the method of transfer learning of convolutional neural network (CNN), and the classification accuracy of the final model is 94.28%. Both of the two models have achieved good performance at last, which proves the feasibility of this method. At the same time, it provides a new idea for further expanding the application of machine learning in automatic fruit classification.

The project aims at avoiding the loss caused by strawberry corrosion in the transportation process, reducing the risk to food safety and the cost and error of manual strawberry grading. A fast and accurate strawberry grading model is trained by building a model and training with a data set.

A computer vision system comprises an LED light-supplementing studio, an object stage, a fixed bracket, a mobile phone, and a computer. The experimental equipment is low in price and cost, convenient to use, simple in operation, and capable of collecting strawberry images well.

When constructing the traditional machine learning model, we conduct the digital processing for the strawberry image in the data set using the RGB color model and get the representation in R, G, and B components in the strawberry image. It is convenient for us to transform the RGB color model into the HSV model and Lab model to extract the color feature of the strawberry.

We had a problem with partitioning the data set because there are many ways to partition the data set: the hold-out, cross-validation, and Bootstrap method. The hold-out method was used to divide the strawberry data set after our team members communicated and discussed it under our tutors' guidance. According to the ratio of 8: 2, we divided the data set into a training set and a test set. The training set was used to generate the model, and the test set was used to evaluate the accuracy and error of the model.

We needed to analyze the strawberry images when building the traditional machine model. We finally adopted the SVM model through online searches and various channels.

Next, we built a deep learning model and prepared to evaluate the accuracy of the deep learning model and compare it with the accuracy of the traditional machine learning model. The convolutional neural networks (CNN) used in this work include the convolution, nonlinear, pooling, and up-sampling layers. Firstly, the RGB color model was used to segment the strawberry image in the data set. And then, we performed convolutional operations based on the convolutional kernel in the convolutional neural networks (CNN) system to realize the extraction of the strawberry image characteristics. This method differed from the traditional machine learning method, which covered all the features of the strawberry image, not only the color features. The acquired strawberry images were preprocessed accordingly and modified the image resolution modified to 224 × 224.

When dividing the data set, we discussed each other and expanded it to 11,340 pieces by image enhancement. The strawberry images were randomly divided into the training set, the test set, and the verification set according to the ratio of 6: 2: 2. Finally, the VGG16 network model was trained in advance on the ImageNet data set for testing.

The following is a schematic diagram of the confusion matrix of deep learning. The horizontal coordinate is the prediction category, and the vertical coordinate is the actual category. We randomly select 594 pictures from each category to evaluate the model's accuracy. It is concluded that the total classification accuracy of this test set is 94.28% (Fig. 7.5).

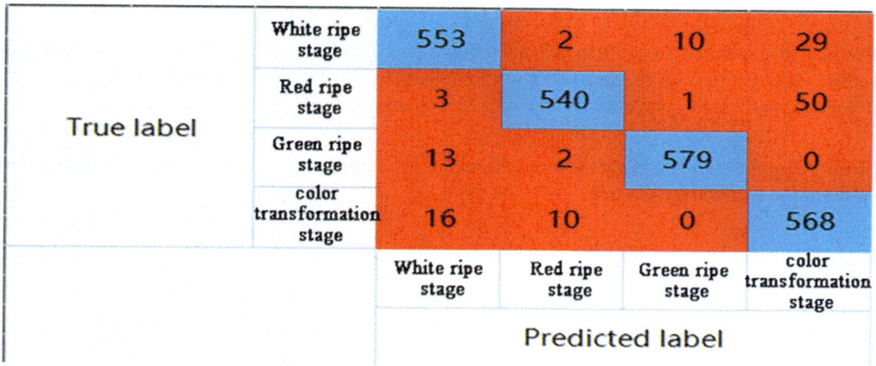

Fig. 7.5 Schematic diagram of deep learning confusion matrix

In the traditional machine learning model, SVM's classification effect is good, with an accuracy of 97.50%. In the deep learning model, we used the learning method of convolutional neural network migration, and the classification precision was 94.28%. In summary, the two models achieved good performance and provided a new idea to expand further the application of machine learning in fruit grading. This project not only adopted the method of building the traditional machine learning model but also tried to use different methods to solve the problem of strawberry grading by building a deep learning model, and the effect was excellent.

Bibliography

Hossain, M. S., Al-Hammadi, M., & Muhammad, G. (2019). Automatic fruit classification using deep learning for industrial applications. *IEEE Transactions on Industrial Informatics, 15*(2), 1027–1034. https://doi.org/10.1109/TII.2018.2875149

Jerome, J., & Vinoth, P. (2017). Artificial intelligent system for measurement of harmonic powers. *ASEAN Journal on Science and Technology for Development, 25*(1), 47–59. https://doi.org/10.29037/ajstd.230

Li, Z., Li, F., Zhu, L., & Yue, J. (2020). Vegetable recognition and classification based on improved VGG deep learning network model. *International Journal of Computational Intelligence Systems, 13*(1), 559. https://doi.org/10.2991/ijcis.d.200425.001

Ribeiro, M. B., Abe, E. S., Kondo, A., Safatle-Ribeiro, A. V., Pereira, M. A., Zilberstein, B., & Ribeiro, U. (2022). Gastric cancer with concurrent pancreatic schwannoma: A case report. *World Journal of Gastrointestinal Pathophysiology, 13*(3), 107–113. https://doi.org/10.4291/wjgp.v13.i3.107

Zawbaa, H. M., Abbass, M., Hazman, M., & Hassenian, A. E. (2014). Automatic fruit image recognition system based on shape and color features. In A. E. Hassanien, M. F. Tolba, & A. Taher Azar (Eds.), *Advanced machine learning technologies and applications* (pp. 278–290). Springer International Publishing. https://doi.org/10.1007/978-3-319-13461-1_27

Zawbaa, H. M., Hazman, M., Abbass, M., & Hassanien, A. E. (2014). Automatic fruit classification using random forest algorithm. In *2014 14th International Conference on Hybrid Intelligent Systems* (pp. 164–168). https://doi.org/10.1109/HIS.2014.7086191

Yuxia Lin graduated from Fuzhou University with a master's degree and is now a junior high school labor technology teacher and guides students to participate in various scientific and technological innovation competitions and achieve certain results.

Siming Huang graduated from Shaanxi Normal University and now works as a junior high school information technology teacher.

Chapter 8
Developing an Interactive Show-Wall with Smart Home Technology for STEM Learning

Xiaohuan Hu, Haosen Xu, Xiayi Lin, Pinxi Jin, Zuoru Xie, and Dongsheng Zhu

Abstract This project is a STEM project jointly built by students, teachers, and enterprises that aim to introduce artificial intelligence elements combining culture and teaching into the decoration design of the lab. The project has great value in inspiring students and serves as a specific reference for other schools to build artificial intelligence laboratories. Due to insufficient ability, traditional AI labs lack personalization and cannot display students' achievements. However, the interactive show-wall with smart home technology reflects the people-oriented characteristics of AI and is closer to daily life. Moreover, the show-wall is a dynamic construction that allows students to use the latest creative works to experiment with AI technology, allowing students to experiment with AI technology, creating a sense of achievement and inspiration for students who participated in the design.

Keywords Interactive show-wall · Smart home · Internet of Things · Open-source hardware

Highlights

This project is a collaborative effort between students, teachers, and enterprises to introduce artificial intelligence elements into the decoration design of the lab.

Zhejiang Wenzhou High School was founded in 1902. In 1953, it was confirmed as one of the 14 key middle schools in Zhejiang Province. In 1981, it was approved as one of the first 18 key middle schools in Zhejiang Province. In 2014, it was awarded the first batch of Zhejiang first-class ordinary high school characteristic demonstration school.

X. Hu · H. Xu · X. Lin · P. Jin · Z. Xie (✉) · D. Zhu (✉)
Wenzhou High School, Wenzhou 325014, Zhejiang, China
e-mail: xiezuoru@vip.qq.com

D. Zhu
e-mail: 290926444@qq.com

X. Hu
e-mail: xh2376@nyu.edu

R. Huang et al. (eds.), *A Collection of AI Innovations by Chinese Teenagers*, Smart Computing and Intelligence, https://doi.org/10.1007/978-981-97-3316-3_8

With this project, students can experiment with AI technology and create a sense of achievement and inspiration.

Challenges Traditional IT labs are generally filled with various instruments but lack sufficient showcasing and exhibition aspects.

Solution We integrated and presented computer vision and Internet of Things (IoT) technologies. With education as the foundation, we prioritized the use of open-source hardware and software, keeping scalability and affordability in mind.

Outcome We developed an interactive AI smart wall that engages users in the Wenzhou dialect and includes intelligent home control terminals and sensors. This allows for facial-controlled lighting, posture-controlled curtains, voice-controlled air conditioning, and pattern-controlled lighting.

What is the purpose of the Interactive Show-Wall project?

The purpose of the Interactive Show-Wall project is to introduce artificial intelligence elements combining culture and teaching into the decoration design of the lab. The project has great value in inspiring students and serves as a specific reference for other schools to build artificial intelligence laboratories.

How does the project incorporate smart home technology?

The Interactive Show-Wall project incorporates smart home technology *by creating an AI interaction wall, which is a collection of intelligent home controllers with human–computer interaction. The data acquisition mainly includes cameras, microphones, lights, acceleration sensors,* etc. *The lab's home is whole-home smart, and an open-source smart home gateway, HASS, integrates intelligent lights of different brands.*

What are the benefits of using an interactive show-wall for STEM learning?

The benefits of using an interactive show-wall for STEM learning are that it allows students to experiment with AI technology, creating a sense of achievement and inspiration for students who participated in the design. Additionally, the interactive show-wall with smart home technology reflects the people-oriented characteristics of AI and is closer to daily life.

8.1 Overview of the Interactive Show-Wall Project

8.1.1 Background

Artificial intelligence education has recently become a hotspot in primary and secondary schools. Many schools with conditions have started to build artificial intelligence laboratories and implement artificial intelligence education. Wenzhou High School is no exception. In 2021, it successfully applied for the first batch of artificial intelligence experimental schools in Wenzhou and completed the construction of basic functions of the laboratory with the support of SenseTime. Besides a high-configuration arithmetic computer server and standard learning terminals, the entire laboratory is also installed with Mijia series smart home devices. As the core members of the school's science and technology production club, we also participated in laboratory construction at the beginning and undertook the laboratory culture construction task under the laboratory instructor's leadership.

Traditional information technology laboratories often place various instruments and equipment and pay little attention to the cultural display. In our opinion, artificial intelligence is a mysterious subject for most students. If there is no cultural display in the laboratory, giving students or other visitors direct feelings is difficult: Where is the *AI* in the AI lab? What is the difference between this and an ordinary computer room? Therefore, we have specially considered how to display AI works in building an artificial intelligence laboratory. In order not to take up space, we think that we can use the wall space, with the wall as the medium, so that the user can directly communicate and interact with AI, so we have this interactive AI background wall project.

8.2 Project Objectives

Following a comprehensive investigation, it has come to our attention that there is a paucity of AI laboratories within primary and secondary educational institutions. Furthermore, within colleges and universities, the existing AI laboratories are often bereft of the vitality and allure necessary to captivate the student body. These laboratories, characterized by their monotonous white walls adorned only by the static portraits of eminent AI luminaries such as Alan Turing, John McCarthy, Calvin Minsky, and Arthur Samuel, remain unaltered year in and year out. Such a lackluster ambiance hardly serves as an enticing environment for prospective students eager to explore the fascinating realm of artificial intelligence. The pressing question then arises: How can an AI laboratory, in its current state, entice students to immerse themselves in the intricate world of AI and actively engage in its study and practice?

Following extensive deliberations and brainstorming sessions, we have collectively arrived at the following consensus:

(1) **Elevating the Aesthetic Experience**: It is imperative that the AI Lab undergo a transformation, shedding its conventional visage. Instead, it should embrace an environment that not only showcases the fruits of AI innovation but also serves as a wellspring of inspiration for students. The AI Lab should transcend the confines of mere computer screens and venture into the tangible realm of smart homes. Imagine, for instance, interactive demonstrations where AI seamlessly collaborates with smart devices like lights, curtain motors, air conditioning units, and more. Students would be enchanted by this harmonious convergence

of technology and everyday life, beckoning them into the mesmerizing world of AI with a sense of wonder and awe.

(2) **The Art of Remake Learning**: Here, the term "Fu Ke," a translation of "Remake" in English, takes center stage. It alludes to the AI projects and works on display, which should serve as not just a showcase but as a fertile ground for students' growth. These projects must be approachable, learnable, and, most importantly, reproducible by students. They should be attainable milestones within their educational journey, allowing them to actively participate in AI development. Moreover, these works should be dynamic, regularly updated, and easily replaceable to keep pace with the ever-evolving landscape of AI technology. At the culmination of each course or project, outstanding student achievements should adorn the laboratory's halls, a testament to their hard-earned skills and innovation.

(3) **Optimizing Spatial Utilization**: We recognize that our school's AI laboratory operates within a finite space, approximately 90 square meters in size—expansive, yet constrained when compared to the sprawling exhibition halls of a science and technology museum. To maximize efficiency and minimize spatial intrusion, the AI laboratory's primary display area should be focused on the background wall. This strategic placement ensures that the limited space is employed judiciously, allowing the essence of AI innovation to shine without overwhelming the physical environment.

In conclusion, by incorporating these transformative concepts and embracing the dynamic potential of AI, we aim to rejuvenate our AI laboratory, making it a hub of creativity, inspiration, and innovation. Through intuitive experiences, hands-on learning opportunities, and optimized spatial design, we aspire to captivate the hearts and minds of students, drawing them into the exciting world of artificial intelligence. The AI laboratory of tomorrow will not merely be a room with white walls and static portraits but a dynamic incubator of talent, where the possibilities of AI are limited only by the imagination and ambition of the students who walk through its doors.

Soon, we will focus on creating an AI background wall to interact with the users.

In our opinion, this AI background wall is the core of the whole AI laboratory and should have three functions:

(1) **Static Function**: The background wall shall have AI-related patterns, such as the words representing the laboratory AI culture and some AI professional terms. These patterns, words, laboratory tables, chairs, and other equipment together form the overall cultural atmosphere of the laboratory.

(2) **Dynamic Function**: The background wall should be variable along with the different applications, teaching scenes, and different cultures. Because there are many unique terms for artificial intelligence, it is not easy to display all of them on a wall, so this wall should be able to be changed and decorated at any time. For example, when you take an intelligent online course, you can change it into keywords related to intelligent online; when you take a computer vision course, you can change it into keywords related to image detection.

(3) **Interaction Function**: AI works can be fixed on the background wall, or AI perception devices can be fixed and then interact with the smart home of the laboratory through these devices. The standard *smart home* uses a central control screen to control the entire home's smart home, and we can use an entire wall. The display must be fantastic if it is easy to fix. Imagine that every time students enter the laboratory, they cannot help interacting with artificial intelligence, so their learning and entertainment will be thoroughly combined.

For ease of presentation, the interactive AI background wall item will be called *interactive wall* from now on.

8.3 Implementation with Smart Home Technology

In the crisp September of 2021, our eager footsteps crossed the threshold of Wenzhou High School, embarking on a new and exhilarating chapter of our academic journey. The corridors of our school were abuzz with whispers of an exciting endeavor underway: our school's ambitious pursuit of becoming a recognized Wenzhou Artificial Intelligence Base School, coupled with the grand vision of establishing a state-of-the-art Artificial Intelligence Laboratory. It was an electrifying time indeed, and our hearts swelled with anticipation.

During this pivotal moment in our academic lives, we found ourselves in the capable hands of two exceptional senior students, Hu Xiaohuan and Xie Ji, who were skillfully guided by the experienced leadership of Xie Zuoru and Zhu Dongsheng. Under their mentorship, we were initiated into the captivating world of artificial intelligence and the remarkable potential it held for our futures.

These two remarkable seniors were quick to acquaint us with the remarkable work undertaken by the Science and Technology Production Club. Their focus, in particular, had been on the fascinating realm of the Internet of Things (IoT). One standout revelation was the Home Assistant (HASS) platform, a veritable marvel that could harmoniously unite a diverse array of smart home products under one virtual roof. What set HASS apart was its ingenious fusion with AI technology, opening the door to the enchanting world of AIoT, or Intelligent Networking—a domain brimming with possibilities and promise.

As the summer vacation of 2022 dawned, some of our senior compatriots embarked on journeys of academic enrichment. For instance, Senior Hu Xiaohuan had the privilege of journeying to the hallowed halls of New York University, while Senior Qiu Yisheng set out for the prestigious East China Normal University. Undeterred by the miles that separated us, we, the remaining members of the Science and Technology Production Club, persevered in our research endeavors under the sage guidance of our beloved teacher, Xie Zuoru.

Our collective focus pivoted toward the intricate task of deploying AI models on compact, open-source hardware designed to be affixed to walls. This innovative approach brought us face to face with the burgeoning realm of edge computing and model compression—a terrain that demanded our unwavering commitment to mastery. It must be noted that these topics were not readily abundant in the vast expanse of the internet. Nevertheless, our thirst for knowledge remained unquenchable, and we tenaciously pursued our quest for understanding.

As the days turned into weeks and our exploration deepened, we found ourselves in the midst of a remarkable transformation. Our initial prototypes, once modest and unassuming, were steadily evolving into fully fledged, display-ready products that embodied our relentless pursuit of excellence. With each passing day, we inched closer to realizing our visionary dream—a cutting-edge AI laboratory that would stand as a beacon of technological innovation and an incubator for future generations of innovators.

In conclusion, our journey through the corridors of Wenzhou High School has been a testament to the power of curiosity, dedication, and the unyielding spirit of exploration. As we stand on the precipice of the future, our hearts are alight with the promise of technological advancement and the boundless horizons that artificial intelligence holds for us. The narrative of our AI laboratory, still in its infancy, is a tale of relentless pursuit, innovation, and the unwavering belief that the world of AI has no bounds, only infinite possibilities.

8.3.1 Architecture Design

In the intricate landscape of AI interaction, we find ourselves traversing a multi-faceted terrain that comprises three essential components: data acquisition, intelligent processing, and control output. These components synergize to create a seamless interaction, orchestrating a symphony of technological marvels. In our quest to bring this interaction to life, we have strategically chosen to concentrate our efforts on data acquisition and intelligent processing. This strategic focus not only streamlines our approach but also serves to curtail the implementation costs associated with constructing the interactive wall.

The third element in this dynamic equation, the control output, takes shape as a diverse array of smart home devices adorning the laboratory's futuristic landscape. These devices encompass a wide spectrum of smart technology, from intelligent lamps that effortlessly illuminate our path to sockets that are imbued with intelligence, responding to our every command. These devices constitute the tangible manifestations of our AI interaction, infusing our laboratory with a palpable sense of technological wizardry.

To visualize the intricate interplay of these components, we have encapsulated our main workflow in Fig. 8.1, a visual testament to the complexity and elegance of our AI interaction system. This diagram serves as a map, guiding us through the intricate labyrinth of data acquisition, intelligent processing, and control output. It embodies our commitment to fostering a harmonious synergy between these elements, culminating in an AI laboratory that is not only cutting-edge but also cost-effective in its implementation.

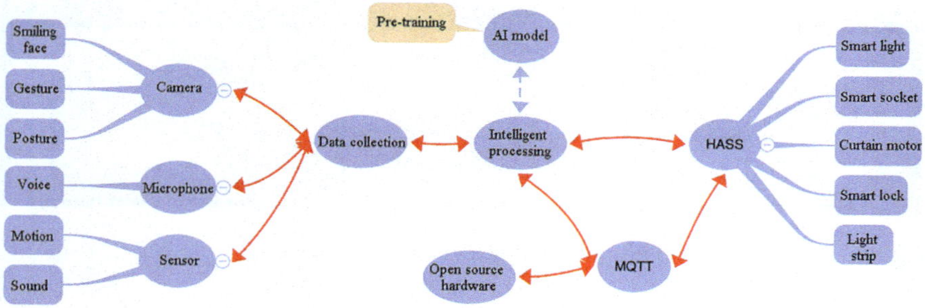

Fig. 8.1 Workflow of AI interactive wall

The AI interaction wall represents a captivating fusion of human–computer interaction and intelligent home control, a dynamic amalgamation of technology and user engagement. Its functionality hinges on a sophisticated network of data acquisition devices, which encompass a versatile range of sensors including cameras, microphones, lights, and acceleration sensors, among others. These sensors serve as the eyes, ears, and touchpoints through which the AI interaction wall perceives and comprehends its surroundings.

It is worth noting that our laboratory has fully embraced the concept of whole-home smart integration, as vividly illustrated in Fig. 8.2. This holistic approach ensures that every corner of the lab is infused with smart technology, creating an immersive environment where innovation thrives. To seamlessly integrate a variety of intelligent lighting solutions from different brands, we have harnessed the power of an open-source smart home gateway known as Home Assistant, affectionately abbreviated as HASS. HASS acts as the linchpin, effortlessly harmonizing the diverse spectrum of smart lighting options, allowing for a unified and intuitive control experience.

In our pursuit of a comprehensive AI interaction ecosystem, we have recognized the importance of remote control capabilities. To this end, we have strategically adopted the MQTT protocol, a lightweight and efficient messaging protocol ideally suited for the Internet of Things (IoT). Leveraging this protocol, we have established an MQTT intranet server, facilitated by SIoT software. This server acts as the nerve center of our AI interaction system, facilitating seamless communication between devices and enabling remote operations that enhance the overall user experience.

In essence, our AI interaction wall is a testament to our commitment to innovation and technological integration. It stands as a beacon of human–computer interaction, where data acquisition, intelligent processing, and control output converge to create a harmonious symphony of smart technology. As we continue to explore the possibilities of AI and IoT, we look forward to the boundless horizons that await, as we chart a course toward a future where technology enriches and empowers our lives in ways previously unimaginable.

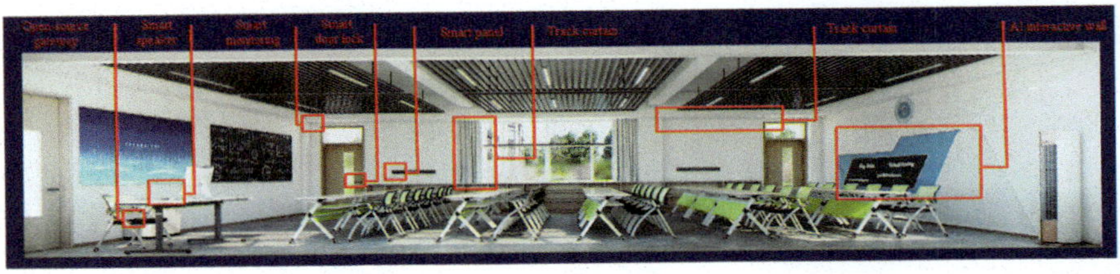

Fig. 8.2 Effect diagram of the intelligent whole-room of AI laboratory

8.3.2 Technical Implementation

8.3.2.1 The Construction of Basic Environment

Installation and Setting of HASS Server

HASS is an open-source project that can run on many kinds of hardware. We chose to run the HASS platform with Raspberry Pi 4 and manage smart devices in the school's AI lab.

HASS provides a variety of installation schemes. Through comparison, the Docker installation scheme is the most convenient. We strongly recommend installing Portainer, a Docker graphical tool, and then installing HASS from the browser.

Enter the following code at the command line to complete the installation:

```
docker pull homeassistant/armhf-hassio-supervisor:latest
docker pull homeassistant/armhf-homeassistant:latest
docker run -d\
--name="home-assistant"\
-v/home/pi/homeassistant:/config\
-v/etc./localtime:/etc./localtime:ro\
--net=host\
--restart=always\
homeassistant/armhf-homeassistant
```

After the Docker container is started, enter IP: 8123(default port number) of HASS in the browser to access the front-end page of HASS. For the first time, you need to wait a while to complete the networking update and then finish the initial setting of the user.

After entering the system, you need to request a long-term access token as the identity authentication of the programmed connection. First, select the user in the bottom column of the page's left sidebar, and then click Create Token at the bottom of the right page. After the creation, please be sure to save the Token's contents. Since tokens vary in content, your token is replaced below.

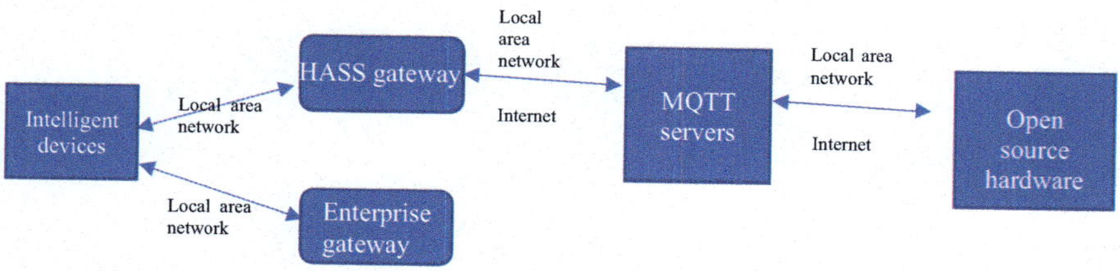

Fig. 8.3 Workflow chart of using open-source hardware to control smart homes

Installation and Setting of MQTT Server

Home Assistant has many extension plug-ins. In addition to managing and controlling smart devices of all brands, Home Assistant also supports the MQTT protocol. Although open-source hardware, such as the mPython, cannot be directly connected to the HASS, it can control the intelligent devices after being transferred by the MQTT server. The operating principle is shown in Fig. 8.3.

It can also be seen from the figure that if we connect the mPython to the MQTT server, the messaging is no longer limited to the Local Area Network, and the intelligent devices in the lab can be implemented anywhere when there is a network.

There are many MQTT servers, and HASS also supports most servers. SIoT is a cross-platform MQTT software suitable for teachers and students in primary and secondary schools. We take it as an example to introduce the construction of an MQTT server. The corresponding version of SIoT software can be downloaded at https://gitee.com/vvlink/SIoT. The software can be run without installation and registration. The default username is *siot*, and the password is *dfrobot*.

The SIoT service runs on port 1883, and the Web management port is 8080, which can be viewed through a browser, as shown in Fig. 8.4.

8.3.2.2 Core Technology Implementation

Realization of Gesture Recognition

Compared a variety of techniques, we finally decided to use the MediaPipe library for gesture recognition, but it is a more complex task to use the MediaPipe library for behavior detection; if this is done, the accuracy of behavior detection entirely depends on the detection accuracy of the critical points of the MediaPipe library. Therefore, first, define the P function to process the image and use WaitTime to the realize sleep state. The code is as follows:

```
# Annotate the joint nodes when a figure is recognized
ifresults.pose_landmarks:
```

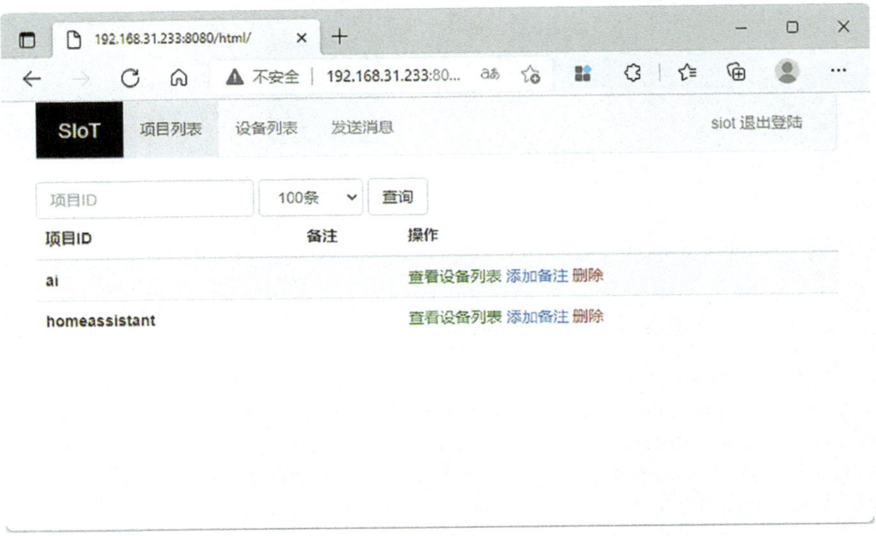

Fig. 8.4 Web administration interface for SIoT

```
mp_drawing.draw_landmarks(img,results.pose_landmarks,mp_pose.POSE_
CONNECTIONS)
for i in range(33):
    cx=int(results.pose_landmarks.landmark[i].x*w)
    cy=int(results.pose_landmarks.landmark[i].y*h)
    keypoints[i]=(cx,cy)
    x[i]=cx
    y[i]=cy
#Draw the figure nodes diagram for easy observation by students and
adjustment of posture to meet the specified poses.
for i in range(33):
    (cx,cy)=keypoints[i]
    img=cv2.circle(img,(cx,cy),radius[i],colors[i],−1)
        cv2.putText(img,"FPS{}".format(str(fps)),(12,100),cv2.FONT_
HERSHEY_SIMPLEX,tl/3,(255,255,0),thickness=tf)
```

Then, according to the critical points of the human body in the following figure, the human joint angle can be calculated to detect the posture of the human body. As shown in Fig. 8.5, the angle between the arm and the horizontal must be greater than 0 degrees when raising the hands.

The definition function returns the human body joint angle. First, acquire the body key points in a single frame, then recall the NumPy library to calculate the angle

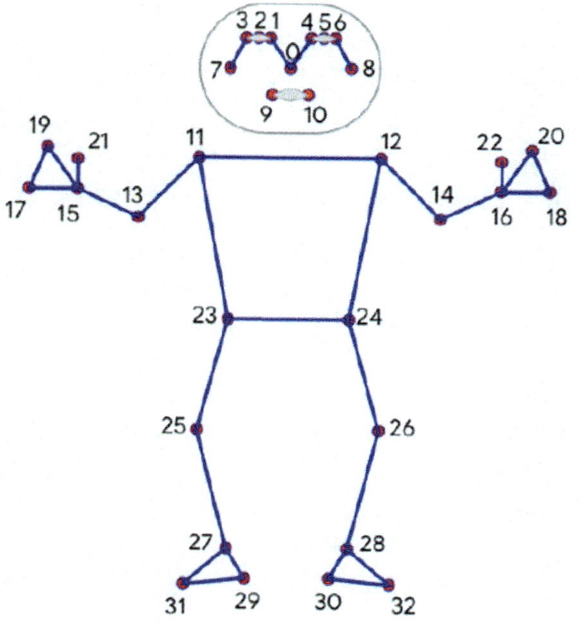

Fig. 8.5 Key points diagram of the human body

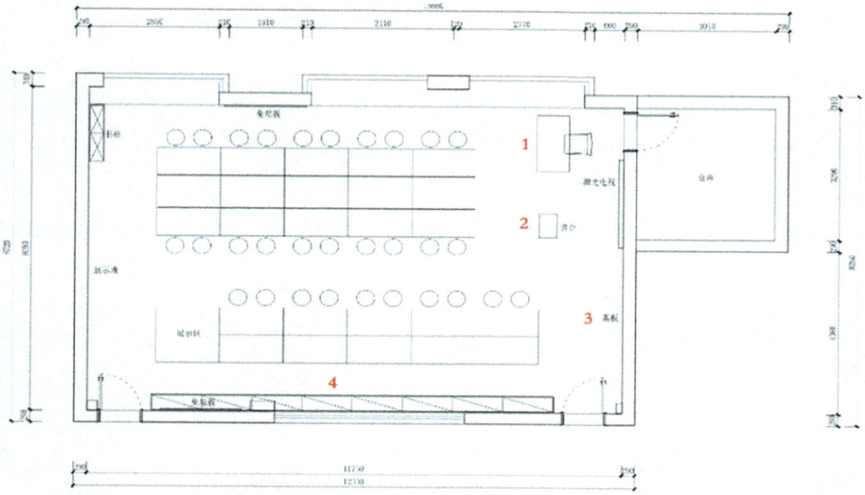

Fig. 8.6 Design of laboratory

between the limbs. Obtaining a vector formula from the coordinate is subtracting two coordinates. The code is as follows:

```
defget_angle(v1,v2):
angle=np.dot(v1,v2)/(np.sqrt(np.sum(v1*v1))*np.sqrt(np.sum(v2*v2)))
angle=np.arccos(angle)/3.14*180
cross=v2[0] * v1−v2*v1[0]
if cross<0:
    angle= −angle
    return angle
```

Design of Control Mechanism

Design a wake-up screen similar to Mi AI, Siri, and Tmall Elf's wake-up words. The user first makes a specific gesture, and the controller sends a warning tone (voice) to indicate that the user has entered the recognition state.

```
# Open the Camera
camera=cv2.VideoCapture(0)
cv2.namedWindow('Cam',cv2.WINDOW_NORMAL)
```

To ensure that the device does not cause the curtain to open or close when we do not need it, we designed a unique wake-up action with the following code:

```
if angle_left_arm< 0 && angle_right_arm> 0 && abs(angle_left_elow)<120
&& abs(angle_right_elow)<120: return
```

After wake-up, we can use the following code to achieve the loop for gesture judgment on the captured image.

```
while True:
success,frame = camera.read()
if success==0:
    print('off')
continue
img=p(frame)
cv2.imshow('Cam',frame)
if cv2.waitKey(125)!= −1:
break
```

8.3.3 *Function Division of Interactive Background Wall*

We chose the wall behind the lab as the main body of the interactive wall. When you enter the AI lab, you look up and see this wall. The overall layout of the laboratory is shown in Fig. 8.6, which is marked as a display wall. It is now the interactive AI background wall. The interactive AI Background Wall has three functions: static, dynamic, and interactive.

8.3.3.1 Static Function

Among the facets of our AI interaction wall, the static function emerges as the most straightforward to bring to life. To embark on this creative journey, we collaborated with a dedicated advertising agency, entrusting them with the task of transforming our laboratory walls into canvases of innovation. With nimble hands and artistic prowess, we opted for the tactile approach of hand-painting the walls using acrylic paint—a medium that seamlessly melds artistry and technology.

The canvas chosen for our artistic endeavor was drenched in a soothing blue-green hue, carefully selected to harmonize with the overall color scheme of our laboratory's furniture and fixtures. This color palette imbues the space with a sense of serenity and freshness, creating an environment conducive to creativity and exploration.

On the left side of these newly adorned walls, a magnificent laboratory LOGO comes to life, etched into existence from an acrylic plate. This emblematic symbol serves as a beacon, proudly proclaiming the unique identity and character of our laboratory. It stands as a testament to our commitment to innovation and serves as a visual anchor, drawing all who enter into the heart of our technological wonderland.

As we continue to celebrate the union of art and science, two essential slogans grace the walls, their significance echoing through the corridors of our laboratory. "Everything can be computed" resonates with the promise of limitless possibilities that the world of artificial intelligence offers. It is a reminder that the power of computation knows no bounds and that it permeates every facet of our existence. The second slogan, "Algorithms are everywhere," serves as a gentle nudge to our curious minds, urging us to recognize the omnipresent influence of algorithms in our daily lives.

Figure 8.7 captures this harmonious fusion of art, technology, and inspiration, encapsulating the essence of our AI interaction wall. It is a tableau that invites all who enter to embark on a journey of exploration and innovation, where the static function is but one facet of the intricate tapestry of AI's boundless potential. As we gaze upon this mural, we are reminded that the future unfolds before us, and within these walls, we are poised to shape it with our unbridled creativity and passion for AI.

The right side of our laboratory wall serves as a canvas for a captivating visual masterpiece—an AI knowledge map that beckons students and enthusiasts alike to embark on a journey of discovery and learning. This intricate map, akin to a roadmap

Fig. 8.7 Static style of background wall

to the realm of artificial intelligence, is a testament to our commitment to fostering a comprehensive AI knowledge system.

The knowledge map is a symphony of AI-related patterns and words, meticulously placed on individual nodes that form an intricate web of interconnected concepts. Each node serves as a beacon, representing a distinct facet of AI knowledge. These patterns and words have been thoughtfully positioned toward the upper part of the wall to ensure their visibility remains unimpeded by the laboratory's tables and chairs, facilitating a seamless and immersive learning experience.

The lines that link these nodes are painted with precision, using the same acrylic paint that graced the rest of our wall. This consistency in design not only maintains a cohesive aesthetic but also underscores the interconnectedness of the AI knowledge landscape. Each line serves as a conduit, bridging one concept to another, allowing students to traverse the intellectual terrain with ease.

The circle nodes, however, represent a striking contrast in both material and technique. Crafted from sturdy iron sheets, these nodes are affixed to the wall with an adhesive super glue, ensuring their permanence and stability. These circular nodes serve as focal points, anchoring the entire knowledge map and drawing the observer's eye into its intricate tapestry.

The result is a visually captivating AI knowledge map, an embodiment of our dedication to cultivating a rich educational environment within our laboratory. As students engage with this mural, they are not merely passive observers but active participants in the process of building their own AI knowledge system. It is a dynamic representation of our commitment to the growth and development of future AI enthusiasts and experts.

In sum, the right side of our laboratory wall is a testament to the fusion of art and education, where the visual splendor of the AI knowledge map complements

Fig. 8.8 Layout of knowledge nodes

the functional utility of the AI interaction wall. Together, they create an environment where knowledge flourishes, curiosity is nurtured, and the endless possibilities of AI are at our fingertips.

8.3.3.2 Dynamic Function

The circle node uses the iron sheet because we want the patterns or texts on the Knowledge Map node to be quickly and dynamically adjusted. According to different courses, hang different words to form a new knowledge map. These knowledge nodes are also easy to make, spray-painted with foam board, then slotted in the back, and embedded with powerful magnets to be absorbed on the iron sheet. The difficulty of production is very low and be made in any advertising store (Fig. 8.9).

8.3.3.3 Interaction Function

The interaction function is the core of the wall, in which the perception (input) part is completed by open-source hardware and camera, the execution (output) part is smart home devices, the AI calculation (processing) part is flexibly deployed according to the complexity of the algorithm, the simple ones are completed directly at the open-source hardware end. The complex ones are implemented with the help of the computing power server.

The round iron sheet on the Knowledge Map node can stick knowledge points and conveniently fix our interactive works. Because the existing open-source hardware is small in volume and light in weight, a line board, a display screen, and a camera are conveniently attracted by a magnet so that visitors can understand the intelligence of artificial intelligence and the convenience brought by the artificial intelligence in the process of interacting with the space board.

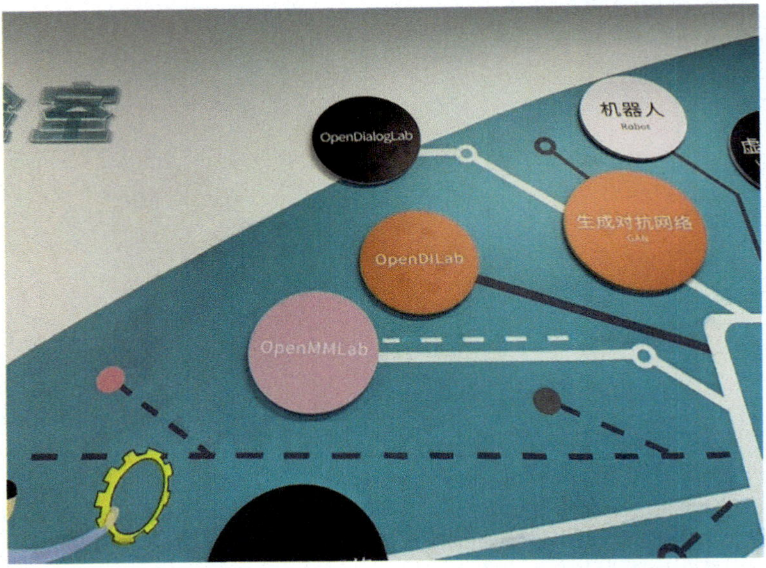

Fig. 8.9 Layout of interactive devices

As shown in Fig. 8.8, we took a mPython, attached our power supply, and taped two magnets to the back. In this way, a magnetic version of the mPython is retrofitted. We can easily mount it on any disc.

It is important to emphasize that the output device of the AI Interactive Wall is the smart home device instead of modules such as LEDs, buzzers, small fans, and steering gear that are common in maker spaces. Because Wenzhou Education Bureau attaches great importance to maker education, many primary and junior school students have been exposed to these standard actuators. We hope that students can use the Internet of Things, AI technology, and products used at home to be closely related to solving real problems. Therefore, the electrical appliances in the laboratory are smart devices, mainly based on Xiaomi's Mijia series, including part of Tmall Elf and HomeKit series products.

8.4 Conclusion

The heart of artificial intelligence, with its enigmatic algorithms and hidden intelligence, presents a unique challenge when it comes to creating tangible, immersive learning environments. Traditional artificial intelligence laboratories, more often than not, resemble ordinary computer rooms, lacking the distinctive features that could make AI come alive for students. While some institutions attempt to showcase AI work displays, the practical limitations of resources and expertise often lead them to feature works exclusively developed by major corporations. This well-intentioned effort, however, inadvertently results in a lack of personalization and an uninspiring atmosphere that fails to capture the imagination of students and educators alike.

In stark contrast, our approach to the AI interaction wall challenges this status quo. Rather than settling for the ordinary, we have engineered a dynamic and human-centric space that bridges the gap between AI technology and our everyday lives. By seamlessly integrating with smart home devices, our interactive wall fosters a tangible connection between artificial intelligence and practical applications, making the subject more relatable and engaging.

One of the most remarkable aspects of our interactive wall is its dynamic nature. As students grapple with the complexities of emerging AI technologies, they have the unique opportunity to contribute their own creative works to replace older pieces on the wall. This dynamic feature transforms the interactive wall into a living, breathing canvas for students to experiment with AI technology, cultivating a sense of ownership and empowerment that transcends traditional educational paradigms.

Furthermore, our project serves as a blueprint for constructing AI laboratories in schools nationwide. Historically, many school laboratory construction endeavors have been fully outsourced, relegating teachers and students to passive user roles. Our innovative approach, however, champions collaboration between educators, students, and industry stakeholders. It embodies a project-based learning model firmly grounded in real-world challenges and scenarios.

In essence, our AI interaction wall is more than just a static installation; it symbolizes a paradigm shift in AI education. It represents the power of collaboration, personalization, and dynamic engagement. As we continue to refine and expand our AI laboratory, we look forward to inspiring and empowering the next generation of AI enthusiasts and innovators. Our vision extends beyond these walls, and we hope to catalyze a movement that reshapes AI education on a broader scale, ensuring that it remains accessible, relevant, and deeply engaging for all.

The works showcased on our interactive wall transcend the realm of mere displays; they possess practical functionality closely tied to the real world. These works, in fact, align seamlessly with the curriculum of our "Entering the World of Intelligent Networks" course, underscoring their pivotal role in advancing the understanding and application of artificial intelligence. Through these works, we not only educate but also inspire, as students interact with tangible AI-driven solutions that have direct relevance to their lives.

The dedicated members of our project team remain committed to an ongoing journey of learning and innovation in the field of AI. We are steadfast in our mission to enrich the AI interaction wall continually. While we focus on optimization, we also place great importance on structuring comprehensive documentation. This documentation encompasses detailed project plans and the technical realization path of our endeavors. Our goal is to shape this project into an open-source, high-quality artificial intelligence education resource tailored for teenagers.

Our vision extends beyond our laboratory's walls. We aspire to ignite the spark of fascination for AI in the hearts and minds of more students, drawing them into the exciting world of AI creation and innovation. As we strive for continuous improvement, we hope to broaden the project's reach, allowing it to transcend the confines of our school and infiltrate educational institutions nationwide. We envision a future where AI interactive walls become a common feature in schools and communities across the country, fostering a deeper understanding of and engagement with artificial intelligence.

In conclusion, our project encapsulates the spirit of education, innovation, and inclusivity. It is a testament to the power of collaboration and a commitment to shaping the future. Through our endeavors, we aim to make AI accessible, engaging, and inspiring for all, leaving an indelible mark on the landscape of AI education in schools and communities far and wide.

Bibliography

Hu, X. (2022). Building an intelligent home control system with Rowan Air Board. *Wireless Communications, 2022*(3).

OpenXLabEdu Help Documentation. (n.d.). https://xedu.readthedocs.io/

Pinpong Python library tutorial documentation. (n.d.). https://pinpong.readthedocs.io/

SIoT Project Manual. (n.d.). https://siot.readthedocs.io/

Xie, Z., & Qiu, Y. (2022a). Building personalized smart home system with home assistant. *China Information Technology Education, 5*, 81–83.

Xie, Z., & Qiu, Y. (2022b). Designing 'AI Magic Wand' with deep learning and Internet of Things technology. *China Information Technology Education, 9*, 77–79+108.

Zuoru Xie Professor of Information Technology. Director of AI Lab at Wenzhou High School in Zhejiang Province, China. Xie has a diverse range of research interests in Interdisciplinary Learning, Maker Education, STEM Education, and AI Education.

Dongsheng Zhu IT teacher. He has led his students to win the first prize at the national level in artificial intelligence competitions.

Chapter 9
An Intelligent Sensible System on Door Locker for Notification of Older People Lock-Forgetting

Hongqi Mao, Kairui Zhang, Yixin Pu, Ligeng Deng, Yunfeng Zhou, and Jia Lai

Abstract The project is to develop an intelligent sensible system for door lockers to notify older people of lock-forgetting, which is safe, intelligent, economical, practical, simple to install, and convenient to guarantee older people's personal and property safety. The device is designed to have detection and alarm functions to remind the elderly to close the door in time. The project focuses on finding the most suitable components through the detection test of different sensors to reduce costs and make the product affordable. The project aims to help the elderly around us to improve their personal and property home safety without changing the existing door lock structure and to bring a sense of security to the elderly's minds.

Keywords Smart · Older · Home · Intelligent

Highlights

This project aims to provide a safe, practical, and affordable solution for older people who tend to forget to lock their doors. With the use of intelligent sensors, this device can detect and remind the elderly to close their doors in time, ensuring their personal and property safety.

Challenges: The aging population often struggles to adapt to new technologies, presenting various daily life challenges. This study seeks to assist the elderly by

---

Chengdu No. 17 High School, established in 1941, is an old public full high school with a moderate scale. It is an urban quality school with high reputation both inside and outside the industry, fine management, and excellent education, and can help students achieve diverse success.

H. Mao · K. Zhang · Y. Pu · L. Deng · Y. Zhou (✉) · J. Lai (✉)
Chengdu No. 17 Middle School, Sichuan 610065, China
e-mail: 763744269@qq.com

J. Lai
e-mail: 364795608@qq.com

R. Huang et al. (eds.), *A Collection of AI Innovations by Chinese Teenagers*, Smart Computing and Intelligence, https://doi.org/10.1007/978-981-97-3316-3_9

incorporating door alarm systems that detect open doors and uncollected keys without altering existing door lock structures.

Solution: A door magnetic switch was used for door status recognition. An infrared sensor was implemented to identify forgotten keys. A microswitch was used to set off a timeout alarm for unclosed doors.

Outcome: We developed a prototype set of smart door locks using open-source hardware.

How does the intelligent sensible system for door locker work?

The intelligent sensible system for door locker is designed to have detection and alarm functions to remind the elderly to close the door in time. The system has two functional modules: one for judging whether the key is inserted into the door, and the other for judging whether the door is closed. The appropriate sensors and electronic modules are combined to realize these functions. When the system detects that the key is not removed or the door is not closed, it will sound an alarm to remind the elderly to take action.

What are the benefits of using this system for older people?

The intelligent sensible system for door locker provides several benefits for older people. Firstly, it helps to ensure their personal and property safety by reminding them to close the door in time. Secondly, it is safe, intelligent, economical, practical, and simple to install, making it a convenient solution for the elderly. Thirdly, it does not require changing the existing door lock structure, which reduces the cost and complexity of installation. Finally, it brings a sense of security to the minds of the elderly and has a specific deterrent effect on wrongdoers.

Can this system be easily installed on existing door locks?

Yes, the intelligent sensible system for door locker is designed to be practical, simple to install, and convenient for the operation of the elderly without changing the existing door lock structure. This means that it can be easily installed on existing door locks without the need for any major modifications. The purpose of the invention is to manufacture an intelligent door lock device that is simple to operate, low in cost, and strong in practicability, which can bring a sense of security to the mind of the elderly and have a specific deterrent effect on wrongdoers.

9.1 Notifying the Lock-Forgetting for Older People Project

9.1.1 Background

In the ever-evolving landscape of modern society, the issue of aging has assumed an increasingly prominent role. The year 2022 saw a pivotal demographic shift in China, where the population of individuals aged 65 and older reached a significant milestone, accounting for a substantial 14% of the total population. This demographic transformation signifies a profound shift in the societal fabric, marking the advent of what can aptly be termed the "aging era."

This demographic transition is a result, in part, of early family planning policies that have led to the geographical separation of many elderly individuals from their children. Consequently, a substantial segment of our elderly population now finds themselves living in solitude, detached from the support networks that familial bonds traditionally provide. The aging process, while natural, brings with it a series of challenges. As the physical health of older individuals deteriorates, bodily functions gradually weaken, and cognitive faculties, including memory and judgment, inevitably decline. These age-related changes manifest in a range of issues, including forgetfulness, which extends to seemingly routine tasks such as securing one's home by locking the door. Unfortunately, this seemingly innocuous forgetfulness has given rise to incidents of stalking, burglary, and, in distressing cases, even robbery. Within our own team, we've witnessed the impact of such incidents on the lives of our members, underscoring the pressing concern of safety accidents affecting older adults.

Moreover, as China grapples with the ongoing challenge of an aging population, an increasing number of individuals are advancing into their golden years. This demographic shift portends a future where the elderly population will continue to swell, heightening the urgency of addressing the complex social issues associated with aging.

In parallel with this demographic shift, we've witnessed a transformation in criminal activity trends. With the widespread adoption of mobile phone payment systems and the comprehensive deployment of surveillance networks, criminal activities have begun to shift their focus from public spaces to indoor environments, prompting a heightened emphasis on home security. In the era before mobile phone payments became ubiquitous, cash transactions were the norm, and crimes such as theft were predominantly committed in public locales. However, the advent of mobile phone payments has led to a decline in cash usage and a correlated reduction in crimes like theft.

Regrettably, some individuals, driven by illicit motivations, have sought alternative avenues for their criminal activities, leading to a surge in home burglaries. Their primary targets often tend to be elderly individuals living alone. Many older adults, influenced by traditional beliefs, continue to keep a stash of cash at home, unwittingly providing would-be criminals with a tantalizing incentive. News reports have increasingly featured accounts of break-ins, burglaries, and thefts, painting a disconcerting picture of security vulnerabilities. It is under these circumstances that the security of one's home, particularly the integrity of the entry point—the door—takes on critical importance. A secure door becomes not just an entryway but the first line of defense in safeguarding both personal safety and property.

In conclusion, the development of a secure, intelligent, and cost-effective door locking system assumes immense significance in the context of aging societies. Such a system can offer a vital layer of protection, ensuring the safety and security of both lives and property for our elderly population. As we navigate the complex challenges presented by an aging society, prioritizing the well-being and security of our elderly citizens is not merely a moral imperative; it is an urgent necessity. By investing in innovative solutions to enhance home security, we can provide our elderly population with the peace of mind and safety they deserve, ultimately fostering a more inclusive, secure, and compassionate society.

9.1.2 Project Objectives

Our endeavor to address the needs of the elderly within our community has led us to a profound and meaningful research direction: the safety and security of our elderly citizens. Following an extensive survey to better understand the specific requirements and concerns of the elderly in our vicinity, we have embarked on a mission to develop an intelligent door lock device that embodies the principles of safety, intelligence, affordability, practicality, ease of installation, and user-friendliness, all while retaining the existing door lock structure. Our ultimate goal is to provide a solution that effectively safeguards the personal safety and property security of the elderly.

The core objective of our research is to create a device that can address a prevalent issue among the elderly—forgetting to close the door properly or take their keys. We aim to achieve this without the need to modify the existing door lock, ensuring a seamless integration with the daily lives of older individuals. By doing so, we seek to offer a cost-effective solution that serves as a timely reminder to the elderly to retrieve their keys and secure their homes. In essence, our invention is designed to instill a sense of security in the minds of the elderly while also acting as a specific deterrent against potential wrongdoers.

Our research pursuits revolve around two primary objectives:

(1) **Development of a Detection and Alarm System**: We aim to manufacture a product equipped with state-of-the-art detection and alarm functions. For instance, the device will be capable of detecting whether the door has been left open for an extended period. If the door remains ajar, it will trigger an alarm, thereby reminding elderly residents to promptly close and secure the door. This feature is intended to prevent unauthorized access and enhance personal safety.

(2) **Sensor Evaluation and Optimization**: We understand the importance of cost-effectiveness in making this solution accessible to a broader demographic of elderly individuals. To achieve this, we will conduct rigorous testing of

different sensors to identify the most suitable components. By selecting sensors that balance cost-effectiveness with high-quality performance, we can create a product that boasts both affordability and reliability. This approach ensures that our solution remains within reach for a wider range of elderly individuals, thereby extending its potential impact and making safety accessible to more people.

In conclusion, our research journey is rooted in a deep commitment to improving the lives of the elderly in our community. By addressing their safety concerns and crafting an intelligent door lock device that meets their specific needs, we aim to bring peace of mind to older citizens and create a safer environment for them to thrive. Our focus on cost-effective solutions underscores our dedication to making a positive impact on the lives of as many older individuals as possible, ultimately fostering a more secure and caring society for all.

9.2 Prototyping an Intelligent Sensible System on Door Locker

From January 2022 to August 2022, we experienced a process from the project's initial conception to the model machine's completion. We have both gained and had lessons and feel we have grown up a lot. The project includes the following phases.

9.2.1 Research Methods

The main research methods we adopt are the investigation, experiment, and literature research methods. Investigation method is one of the most used methods in scientific research. It is a purposeful, planned, and systematic collection of material about the actual or historical state of the subject.

The members of our group divided the work and investigated at least ten older people in their respective communities about their views on the traditional door lock. Some of the older people still use old-fashioned locks. Some of them live with their children, and they have replaced fingerprint locks. However, they all used old-fashioned locks and experienced years of doors opening with keys, so their views on old-fashioned doors are compelling. We asked them about their concerns about real-world security in the matter of door locks and found that their main problems were that they often forgot to pull out the key and close the door, and they feared that their personal and property safety would be threatened. Based on our survey of the elderly, we have determined that the direction of the study will be around the two points of pulling out the key and closing the door. In the process of the survey, because the older people's eyesight is not very good, and they cannot use their mobile phones to do the multiple-choice questions, so we gave up the paper questionnaire and the questionnaire on their mobile phones and used the chat way to carry on the oral survey to the older people. It has shown that this approach is more acceptable to the elderly, and they are willing to cooperate with our survey.

In addition to investigating the actual needs of the elderly, we also surveyed the existing mature smart door lock products. We find that most of the current intelligent door lock products have the following functions: doorbell function, fingerprint unlocking, password protection, information management, alarm, remote unlocking, voice prompt, etc. Among them, the functions of fingerprint unlocking, password protection, information management, and remote unlocking all involve changing the structure of the existing lock, which does not conform to our original intention, and our knowledge and technology cannot be achieved. Therefore, in combination with the actual situation and according to our existing knowledge and technology, we expect the alarm function to be mainly realized based on not changing the existing door lock structure to detect whether the key is pulled out and the door is closed correctly.

Experimental method: Experimental method is a scientific research method to discover and confirm the causal relationship between things by changing the principal branch and controlling the research object. Around the problem of whether the door is closed or not and whether the key is pulled out, we found online that a reed switch, microswitch, infrared correlation sensor, range sensor, ultrasonic sensor, and laser sensor may be helpful for our project. Our team bought these sensors separately and tested them in turn. We found that reed switches, microswitches, and range sensors could be considered to detect whether the door was closed. An infrared correlation sensor, ultrasonic sensor, and laser sensor may be considered to detect whether the key has been removed. Through our repeated experiments and in combination with the parameters of these sensors given by the seller and based on comprehensively considering the cost price and experimental data, we think that the two components of the microswitch and infrared correlation sensor are more suitable for our requirements as our final choice.

Literature research method: The literature research method mainly refers to collecting, identifying, and sorting out literature and forming a scientific understanding of facts through the literature study.

Since our knowledge and time are very limited, our learning comes mainly from the Internet. To understand the sensors' knowledge, we mostly learn from Baidu Encyclopedia and technical materials provided by Taobao sellers. When searching the case, we mainly from the Baidu library and the China National Knowledge Infrastructure (CNKI) to search for similar literature. Many technical terms are still esoteric and incomprehensible today, but other people's cases can also inspire us to optimize our products. At the end of the report, we will list the relevant literature.

Applying the above research methods, we completed our project in practice stumblingly. We also found that these research methods are adequate for the project and helpful for our study and life.

9.2.2 *Requirement Analysis and Preliminary Design*

The journey of our research team, which commenced in February 2022, has been marked by thoughtful deliberation and a commitment to addressing the pressing concerns of our elderly community members. At the outset, we embarked on a collective exploration of potential projects, with two distinct ideas at the forefront of our discussions: the development of a clothes dryer and the creation of an intelligent door lock. Through a process of rigorous assessment and analysis, we endeavored to determine which project would best align with our mission to enhance the safety and well-being of older individuals.

Our initial exploration led us to examine the working principles and cost considerations of a clothes dryer, drawing from the insights provided by the Baidu encyclopedia. We also sought inspiration for the dryer's design on Taobao. It quickly became apparent that while the principle of a clothes dryer is relatively straightforward, its material composition is intricate and multifaceted. Moreover, we recognized that the level of intelligence inherent to such a device is limited, offering little room for innovation. After careful consideration, we collectively decided to set aside the idea of pursuing a clothes dryer project and instead focused our attention on the more promising and impactful realm of intelligent door locks.

To inform our direction, we conducted a comprehensive survey, reaching out to at least ten older individuals within our respective communities. We engaged them in discussions about their experiences with traditional door locks and sought to understand their concerns regarding security issues. A consistent theme emerged from these conversations: the recurring fear of forgetting to remove the key and securely close the door. The potential threat to both personal safety and property security loomed large in their minds.

With these invaluable insights in hand, we crystallized our research subject and direction: the creation of an intelligent door lock system that can effectively verify whether the door has been securely closed and the key has been extracted. Our mission was now clear—to harness technology and innovation to mitigate the concerns of older individuals and provide them with a robust solution that enhances their safety and peace of mind.

In the chapters that followed, our research team embarked on a journey of exploration and invention, fueled by the collective determination to make a meaningful difference in the lives of our elderly community members.

9.2.3 *Prototype Manufacture and Test*

9.2.3.1 Parameter Collection

Our journey toward creating an intelligent door lock system was underpinned by meticulous research and a systematic approach to technology integration. It began with the collection of technical parameters for the sensors we had procured, a crucial step that served as the bedrock for our sensor utilization.

Drawing inspiration from real-life scenarios, we embarked on measurements that would inform the design and placement of our sensor system. We meticulously gauged the dimensions of the key once it had been inserted into the keyhole, determining both its length and width. This data was instrumental in establishing the precise mounting height for our sensor on the door. Additionally, we measured the gap between the door and the door frame, a critical factor in determining whether the door was closed securely. This meticulous data gathering ensured the accuracy and reliability of our system.

Our pursuit of knowledge extended beyond the physical measurements and sensor specifications. We delved into the realm of technical literature, seeking information on various components integral to our project, including sensors, buzzers, horns, and LED lights. We also embarked on a learning journey focused on Arduino control boards and programming, recognizing their pivotal role in our project's implementation.

However, we encountered a challenge along the way. While we were enthusiastic learners, we lacked a foundational understanding of programming. Undeterred, we adopted a pragmatic approach. We meticulously designed a process framework and engaged our teacher's expertise to help us write the necessary program. This collaborative effort allowed us to leverage the knowledge and skills of experienced individuals while ensuring that our project remained on course.

In essence, our research and development process was a testament to our commitment to excellence and innovation. It showcased our ability to combine practical measurements with comprehensive knowledge acquisition, all while surmounting programming obstacles through collaboration and mentorship. This multifaceted approach was pivotal in our journey to create an intelligent door lock system that would ultimately enhance the safety and well-being of our elderly community members.

9.2.3.2 Selection of Sensors

After looking up the knowledge of the sensors, we found that reed switches, microswitches, infrared correlation sensors, distance sensors, ultrasonic sensors, and laser sensors could all help our project, so we bought them for testing.

When testing whether the door is closed, through experiments, we find that the distance sensor requires remarkably high accuracy. When the door is closed, the data sent by the distance sensor is not closed, which will cause our model to alarm continuously. Reed switches require magnet attraction to work, but we cannot control the strength of the magnet's magnetic field. Sometimes the alarm stops when the door is closed, but sometimes the alarm does not stop when the door is closed, or the sensor does not send a warning when the door is open. We could not find the reason, so we gave up the sensor. We finally decided to use microswitches, which must be touched and sent signals. We also need to consider whether to make two sets of control systems or one set of control systems when designing. After research, it is found that judging whether the door is closed and the key is taken down does not conflict. The same main control board adopting time-sharing multiplexing technology can realize these two functions.

We tested the ultrasonic sensor, the laser sensor, and the infrared sensor when testing whether the key was inserted into the door lock. We tested the ultrasonic sensor first. We found that the operating range of the ultrasonic sensor is a sector, and it is impossible to detect the existing obstacles point-to-point. If there is an obstacle in this sector, the sensor may also assume that the key has been inserted. Furthermore, the volume of the ultrasonic sensor is relatively large, which brings us a certain difficulty in installation. Based on this, we gave up the ultrasonic sensor (Fig. 9.1).

In the quest to find the most suitable sensor for our intelligent door lock system, we explored various options, including the laser correlation sensor and the infrared correlation sensor. While both sensors share similarities with their principle of operation, the laser correlation sensor stands out for its exceptional precision. However, as we conducted experiments and delved deeper into the intricacies of our project, we encountered a notable challenge—door vibrations during the opening and closing process.

These subtle vibrations, though seemingly insignificant, exerted a considerable influence on the laser correlation sensor's signal acquisition. The laser sensor's sensitivity to even the slightest movement posed a significant hurdle. Additionally, we observed that the laser contra-reflective sensor, much like the ultrasonic sensor, had

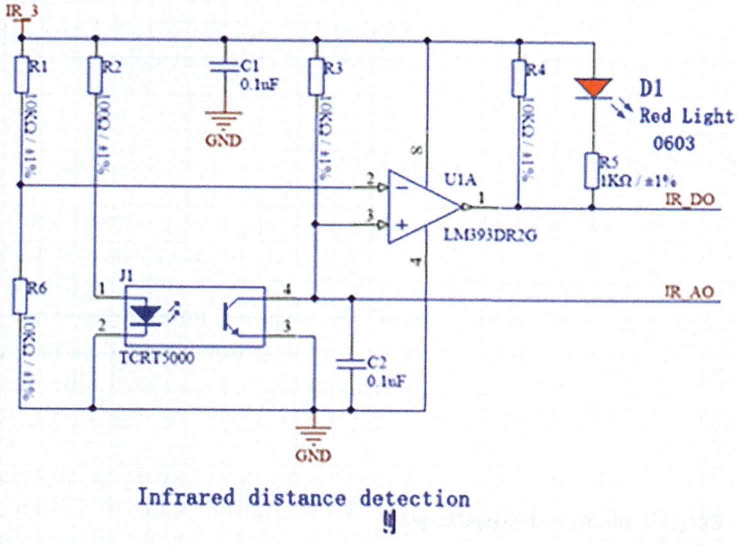

Fig. 9.1 Circuit diagram of infrared distance detection

a relatively bulky form factor. This posed practical challenges during installation and placement.

After careful consideration and experimentation, we ultimately arrived at a decision: the adoption of the infrared correlation sensor. This sensor possessed several compelling advantages that aligned with our project's requirements. Notably, the infrared correlation sensor is compact and easy to install, mitigating the installation difficulties posed by larger sensors. Furthermore, the receiver component of the infrared sensor demonstrated a high level of sensitivity to the signals emitted by the transmitter.

In our finalized system, the infrared correlation sensor plays a pivotal role. It serves as the mechanism for detecting whether the key has been pulled out or not. When the infrared receiver detects the presence of the infrared rays emitted by the transmitter, it signifies that the key has indeed been extracted. Conversely, if the receiver fails to receive the infrared rays, it indicates that the key remains in place, triggering an alarm.

Our alarm system, designed to ensure the utmost safety and security, comprises a combination of LED lamps and a loudspeaker. The LED lights primarily serve as a visual warning mechanism, signaling that the door has not been securely closed. In parallel, the loudspeaker amplifies the auditory component of the alarm, primarily used to indicate that the key has not been removed. This multi-sensory approach enhances the effectiveness of our intelligent door lock system, providing clear and unmistakable signals to users.

In essence, our journey through sensor selection exemplifies our commitment to precision, practicality, and user-friendliness. It underscores our dedication to crafting a system that prioritizes the safety and well-being of older individuals while maintaining ease of use and reliability. The infrared correlation sensor, with its compact form factor and high sensitivity, emerged as the ideal choice to achieve these goals, solidifying its role as the linchpin of our intelligent door lock system.

9.2.3.3 Prototype Testing

To address the issue of doors not being closed in a timely manner and to ensure the safety and security of our intelligent door lock system, we have implemented a well-thought-out mechanism.

For the scenario where the door is not closed on time, we have integrated a microswitch into our system to precisely determine whether the door has been fully closed. The system initiates a timer the moment the door is opened. During this initial phase, if the system's predefined time limit is not exceeded, it simply conducts timing without triggering any alarms.

However, should the duration of the door being open surpass the designated time specified in the system's program, the microswitch remains unactivated, signaling an alarm condition. Simultaneously, the LED lamp begins to flash continuously, serving as a prominent visual alarm indicator to remind individuals inside the premises to promptly close the door. This continuous flashing serves as a clear and unmistakable signal, ensuring that the occupants are aware of the open door.

When the door is finally closed, the microswitch is engaged, and the system's alarm is automatically disarmed. The LED lamp ceases its flashing and returns to a steady state, signifying that the door is now securely closed, and the alarm condition has been resolved.

In the context of addressing the issue of forgetting to remove the key, we have implemented an infrared correlation method to provide a reliable means of key detection. The infrared transmitter and receiver are strategically positioned in alignment with the keyhole. Upon system activation, the infrared transmitter immediately emits a beam of infrared rays, which are received by the infrared receiver.

In the event that the key has not been removed, its presence obstructs the path of the infrared beam, preventing the infrared receiver from receiving the emitted infrared rays. Should the predetermined time interval be exceeded without key removal, the system automatically triggers an alarm. A voice module within the system issues a vocal notification, audibly stating, "Please remove the key," serving as a timely and personalized reminder to promptly extract the key.

This meticulously designed system ensures that both the door closure and key removal processes are closely monitored, with clear and effective alarm mechanisms in place to enhance safety and security. It exemplifies our commitment to creating a comprehensive, user-friendly, and robust intelligent door lock system that meets the unique needs of our elderly community members (Fig. 9.2).

The process of creating our intelligent door lock system was a journey filled with challenges, creativity, and continuous problem-solving. As we ventured into this project, we needed to consider the intricacies of how doors are opened and closed, accounting for variations in each action and the measurement range of each component. Additionally, we had to calculate the required voltage for each element, including the main control board, while taking into account various complications that arose during the development process.

One significant constraint we faced was budget limitations. As a result, we were unable to employ advanced techniques like laser engraving or utilize precision instruments to craft our doors. Instead, we had to rely on manual craftsmanship, building the doors with our own hands. As we progressed in programming and testing, we encountered issues with the wiring, which necessitated drilling holes in the doors for modifications.

In our pursuit of simplicity and user-friendliness, we decided to revisit the wiring design to ensure easy installation. However, this adjustment introduced a new problem with the sensors, leading to a logic error that triggered an alarm when it should not have occurred after the door was closed. The issue of ringing and lighting persisted, prompting us to reevaluate our approach.

To overcome these challenges, we took a step back and revised the system's flow chart. This simplification process ultimately led to a successful resolution of the problem. The collaborative nature of our group was put to the test when we encountered disagreements regarding the placement of the sensors. Some team members argued for sensor installation on the door frame, while others advocated for placement

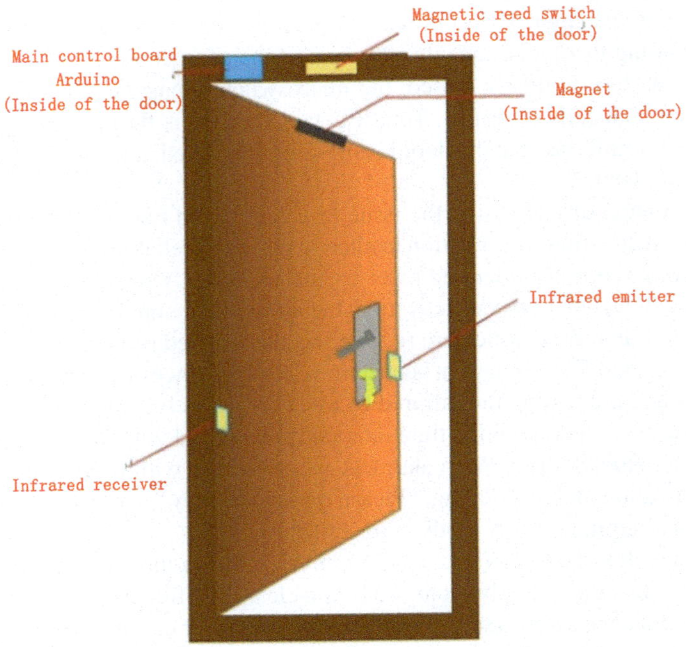

Fig. 9.2 Schematic diagram of detection when the key is forgotten to be pulled out

on the door itself. After thorough discussion and consideration, we concluded that the door frame required longer wire lengths and higher sensor accuracy. In contrast, the door did not demand such stringent requirements, leading us to opt for sensor installation on the door.

The intricate process of connecting each electronic component to the main control board using breadboard wires required meticulous attention to detail. Any misstep in the wiring could potentially result in damage to the component or motherboard, underscoring the need for precision and caution.

Throughout this journey, our team members experienced growth as individuals and as a collective unit. Our discussions sparked intellectual collisions, our passions burned brightly, and our creativity was constantly stimulated. The project not only enriched our technical knowledge but also fostered a sense of camaraderie and determination to overcome obstacles. As we persevered through the challenges, we emerged stronger and more capable, demonstrating the power of teamwork and innovation in the pursuit of a common goal.

9.2.4 Optimization After Tests

The journey of developing our intelligent door lock system didn't end with the basic design and production phases. Instead, we recognized the importance of ongoing optimization to make our product even more effective and user-friendly. With a deep understanding of the daily routines and habits of older individuals, we sought to refine our system by incorporating real-world considerations and introducing the concept of delay.

One key aspect of our optimization was to reconsider the actions of older individuals when they enter and close the door. We wanted to ensure that our system aligned seamlessly with their needs and behaviors. To achieve this, we introduced delay processing to fine-tune our system. For instance, when our system starts its timing process upon the door's opening, we implemented a delay mechanism. If the door remains open beyond the preset time limit, the system triggers an alarm—a delay alarm. This enhancement ensures that older individuals have ample time to complete their tasks before any alarms are sounded.

Furthermore, we took a holistic approach to refining our system. We recognized that it was essential to provide distinct alarms for different scenarios, namely when the door was not closed and when the key had not been pulled out. To achieve this, we introduced new components to our system, allowing us to differentiate between these two conditions.

By controlling these new components, we were able to create distinctive alarm signals for each situation. For instance, the sound produced by the horn alarm was distinct for cases where the door was not closed compared to scenarios where the key had not been removed. Simultaneously, we synchronized the LED lights to correspond with these different alarm states. This visual cue reinforced the auditory alarm, ensuring that older individuals could easily distinguish between the two situations.

In essence, our optimization efforts were driven by a deep commitment to enhancing the usability and effectiveness of our intelligent door lock system. By taking into account the real-life behaviors of older individuals, introducing delay mechanisms, and refining our alarm signals, we aimed to create a system that not only prioritized safety but also catered to the unique needs and experiences of our target users (Fig. 9.3).

The journey of creating our intelligent door lock system was marked not only by successes and improvements but also by the recognition of its shortcomings. While we had worked diligently to refine our model, we acknowledged that there were still areas in need of enhancement, particularly if our system were to be applied in real-life scenarios.

As we progressed through the mid-term of the project, we encountered a challenge that is not uncommon in group endeavors—some team members began to lose enthusiasm due to factors such as heavy academic assignments and time constraints. This dip in motivation had an adverse impact on the project's overall progress and completion. In response to this situation, our group held several meetings, each serving as a platform to rekindle everyone's confidence and commitment to achieving

Fig. 9.3 Prototype of door lock devices

the project's objectives. These discussions provided the necessary encouragement to strive for excellence and ensure that the project reached its full potential.

During the summer break, our entire team participated in a cross-strait computational thinking summer camp for students. This experience served as a valuable opportunity to broaden our horizons and acquire essential project skills. It deepened our understanding of project development and expanded our knowledge of artificial intelligence. Through online courses and interactions with fellow participants, we learned where to access more specialized knowledge in related fields and gained insights into the critical aspects of designing a product.

Furthermore, our exposure to different landscapes during the defense phase of our project was enlightening. We observed that in coastal areas, the level of science and technology was notably advanced. Young individuals, some even younger than us, were already deeply engaged in studying how to harness artificial intelligence to improve and innovate. Witnessing their applications of AI, we learned the importance of foresight and staying ahead of the curve.

Additionally, we gleaned valuable insights during the defense presentations. We discovered the significance of composure while on stage and the teacher's preference for a narrative-style presentation of the project process. Armed with this knowledge, we honed our ability to articulate the project journey effectively, making it not only informative but also engaging.

In essence, our project journey was characterized by a dynamic process of learning, adaptation, and growth. We encountered and addressed challenges, rediscovered our enthusiasm, and expanded our skill set through various experiences. This journey of discovery and development left us better equipped to tackle future projects and contribute meaningfully to the world of artificial intelligence and technology.

9.3 Conclusions and Future Plans

Through this project, we deeply realized that we have many shortcomings compared with others. We opened our eyes wider through this project. By seeking and reading the relevant literature, we know that a product is priced so high but still can be accepted and purchased because it can significantly facilitate people's life and provide convenient services for them. Although our products are not high in technology content, if we can add the service function of our products, they will be accepted by more people.

While doing the project, although we did not have the opportunity and time to learn the standardized programming language, we learned the mode of thinking. This activity made our logical thinking stricter and our thinking manner more diversified. At the same time, it also improved our part of the discipline accomplishment. For example, when comparing different sensors, we can only see the knowledge of sensor-related sensors in books, but we have no chance to practice it. Through this activity, we directly contact the sensor and understand its application method. Real knowledge comes from practice, and not all written books remain unchanged. We feel the gap between experimental and theoretical data.

In this activity, we also learned how to cooperate. Cooperation is essential in human society. It requires the concerted efforts of all members to achieve the goal. If there are differences in the cooperation process, we shall work together to find a solution. When there is a conflict in a team, we must work together to resolve it and reunite. Facing sudden problems, we need to think about how to solve the problem and optimize the solution in each case. Have the spirit of a newborn calf that is not afraid of tigers. For the questions, we should not give up easily. We should go down into the deep water and try our best to solve the problem. In this event, we also learned the importance of innovation. At first, we did not have any ideas about innovation, but after one person put forward an idea, the ideas of innovation came out. Innovation not only comes from technology but also from life. For us, innovation is more of a small step on the shoulders of giants. This activity also enhances our hands-on ability. By making doors and door frames, we learned how to turn two-dimensional drawings into three-dimensional models, making our ideas visible and tangible and exercising our ability to perceive three dimensions. It seems that our ways of thinking have also changed. We have learned to think a few more steps from thinking only about the next step. We have gradually formed forward-looking thinking.

From the early stage of the project to finding the project in life, after we thought about and discussed it, the intelligent door lock project scheme was formed. For the disassembly of the product functions, there are two functional modules, one is for judging whether the key is inserted into the door, and the other is for judging whether the door is closed. And then inquired about the relevant knowledge according to each functional module to find the appropriate sensors and electronic modules; then, the whole manufacturing process: several functions of the intelligent door lock can find the corresponding sensors and electronic modules to be combined to realize. We also reviewed the electrical knowledge and took the initiative to learn new sensor knowledge. In the project, we have found that it is possible to turn ideas into reality if we find a way. In this process, we also discovered our advantages and product advantages and saw many shortcomings.

The product has the advantage that the door is retrofitted based on not changing its shape and function to realize the detective functions whether the key is taken down or not and whether the door is closed. Because the design of the project does not change the basic structure of the door lock, the manufacturing cost is relatively low. The user group is clear, and the project has good practical and economic value.

Our advantages: Good energy, no fear of difficulties, daring to challenge, and creative consciousness. Not much knowledge but willing to learn, good at learning, joint initiative, cooperation, and help, strong hands-on ability.

Shortcomings: Because it is all done manually, the precision of the product is not enough, directly affecting the realization of functions. In addition, the appearance of the product needs to be redesigned because the volume is large and takes up a small amount of space, and so on, which needs to be improved later. Because we are an ordinary high school, the daily quantity of homework is vast. Without a programming foundation, we also encountered great difficulty in production. To realize this project and finish the product prototype within the specified time, we turned to professionals and foreign aid teachers for help. Finally, we finished the product, which realized the basic functions. Therefore, if we want to continue on the road of innovation and creation, we have much to learn.

Although the event is ending, our innovation is still not over. According to our phased discussion, our smart door lock device product can also dig deep into its capabilities by adding electronic modules to:

(1) Communication between the intelligent door lock device and the mobile phone can be added. By adding the WI-FI module, the alarm signal that the door is not closed or the key is not taken off will be linked to the smart phone of the older person. The older person is better reminded to close the door and take the key through the secondary reminding function of the mobile phone.
(2) The automatic door-closing function of an intelligent door lock device can be added. Now there are relatively mature door-closing products on Taobao. We can combine this product with our intelligent door lock device to make the door-closing operation more intelligent.
(3) The function of the doorbell can be added to expand the function of the intelligent door lock device.

All in all, through this project, we sweat and grow. Soon, we will apply the methods and skills learned in the project research process, keep the consciousness of innovation, and adhere to a rigorous work style to grow up better and faster.

Bibliography

Chen, C., Li, J., Liang, Z., & Shi, J. (2021). Collision warning and monitoring smart helmet based on Arduino. *Science, Technology, Innovation and Application, 1*, 53–55.

Li, D., & Wang, J. (2020). Logical design and implementation of gate passage based on multiple infrared sensors. *Railway Communication Signal Engineering Technology, 17*(4), 71–75.

Mou, X. (2021). Dual-channel programming design for Arduino emergency warning device. *Electronic Journal, 006*.

Yunfeng Zhou is teacher at Chengdu No. 17 High School. He has won Ye Deyuan Fund award, 2021 Jinjiang District Teaching Rookie, and 2020 Jinjiang District Teacher Teaching Skills Competition first prize.

Jia Lai graduated from the School of Computer Science of Sichuan Normal University and has been engaged in high school computer education for 17 years.

Chapter 10
Designing a Smart Car for Connecting Home and Community Delivery Pick-Up Point

Liyan Yao, Yaojun Zhang, and Lihuan Wang

Abstract The growth of online shopping and express delivery has resulted in an increased demand for last-mile delivery services, which is still being done through door-to-door delivery in many communities. This mode of delivery is costly in terms of time and resources. The unmanned trolley proposed by smart logistics is an intelligent device that can navigate autonomously and recognize its environment, making it a promising solution for express community delivery. This study analyzes the feasibility and application demands of the smart car in community express delivery and develops a solution that includes features such as path planning, automatic obstacle avoidance, and ultraviolet disinfection. The proposed system can improve delivery efficiency, reduce workforce, material, and time costs, and meet the needs of residents for timely express delivery.

Keywords Smart car · Community express delivery

Highlights

This project explores the feasibility and application demands of using a smart car for community express delivery and proposes a solution that can improve delivery efficiency and reduce costs.

Beijing Sanfan Middle School is the middle school of the Second High School affiliated to Beijing Normal University. The school implements all-round quality education to develop students' core qualities. Establish a diversified curriculum system, adhere to the student-oriented, and actively meet the needs of different students. The school actively explores interdisciplinary courses and other courses to cultivate new people who can adapt to future social development.

L. Yao · Y. Zhang · L. Wang (✉)
Beijing Sanfan Middle School, Beijing 100120, China
e-mail: wanglihuan1106@sina.com

L. Yao
e-mail: wuyu@bnu.edu.cn

Y. Zhang
e-mail: zhangjb@bnu.edu.cn

R. Huang et al. (eds.), *A Collection of AI Innovations by Chinese Teenagers*, Smart Computing and Intelligence, https://doi.org/10.1007/978-981-97-3316-3_10

Challenges: With the rise in online shopping and express delivery services, many packages are delivered daily. This project aims to solve the 'last 500 m' delivery problem within residential areas using mobile delivery lockers that can autonomously move to the resident's doorstep.

Solution: Used line following and ultrasonic obstacle avoidance simulation for navigation in complex environments, and AI technology for road sign recognition. Implemented voice commands and IoT technology to manage the delivery, retrieval, and user acknowledgment of packages.

Outcome: A product prototype set was developed, capable of full functionality in a simple environment.

What are the features of the proposed smart car system for community express delivery?

The proposed smart car system for community express delivery includes features such as route planning and optimization, automatic obstacle avoidance, and ultraviolet disinfection. Additionally, it mentions that before delivery, the system confirms with residents whether they are at home and whether it is convenient for the car to deliver the goods to their homes and notifies the customer to pick up the goods and informs them of the password for taking them to prevent the incorrect goods from being taken and to confirm that the goods have been received.

How can the use of a smart car improve delivery efficiency and reduce costs?

The proposed smart car system for community express delivery can improve delivery efficiency, reduce workforce, material, and time costs, and meet the needs of residents for timely express delivery. This is achieved through features such as route planning and optimization, automatic obstacle avoidance, and ultraviolet disinfection. By automating the delivery process, the system can reduce the need for human labor and increase the speed and accuracy of deliveries. Additionally, the system can optimize delivery routes to reduce travel time and fuel costs and disinfect packages using ultraviolet light to reduce the risk of contamination.

How is this smart car prototype designed, and what specific technologies are utilized in its development?

The smart car prototype was designed based on the ideas of automating the delivery process, optimizing delivery routes, and disinfecting packages using ultraviolet light. The prototype mainly includes a general control module, an execution module, a navigation module, an information communication module, a sense module, a man–machine interaction module, and an autonomous learning module. The smart car prototype integrates a variety of technologies, including a control unit, a navigation module, a driving unit, a road identification and avoidance module, an information communication module, a human–computer interaction module, and more. Additionally, it mentions that the smart car prototype uses a solar panel to charge the car during driving and shutdown automatically, and that the car's components are made of environmentally friendly materials and waterproof as far as possible.

10.1 Overview of the Smart Car Project

10.1.1 Background

In our increasingly interconnected and digitalized world, the rise of online shopping and e-commerce has fundamentally transformed the way we obtain goods and services. Central to this transformation is the dynamic and ever-expanding express delivery industry. Recent data from the State Post Bureau vividly illustrates the impressive growth and scale of this sector, underscoring its significance in our daily lives.

In the year 2021 alone, the total business volume of express service enterprises across China reached a staggering 108.30 billion pieces, representing a remarkable year-on-year growth of 29.9%. This exponential surge in business volume correspondingly translated into substantial revenue, with total business income amounting to 10,332.3 billion RMB, marking a robust year-on-year increase of 17.5%. These figures not only reflect the remarkable growth trajectory of the express delivery industry but also signify its deep integration into the fabric of our society.

It is important to note that express delivery has evolved from being a mere convenience to becoming an essential part of our daily routines. The urban landscape, in particular, has witnessed a significant surge in delivery activity, with a city-wide business volume of 14.11 billion pieces in 2021, showcasing a notable 16% year-on-year growth. Moreover, across various cities, the combined business volume amounted to a staggering 920.8 billion pieces, reflecting an impressive year-on-year growth of 32.8%. Even international deliveries and those to regions such as Hong Kong, Macao, and Taiwan contributed to the thriving industry, with 2.1 billion pieces and a substantial 14.6% year-on-year growth rate.

Despite the undeniable convenience and ubiquity of express delivery, a closer look at community-level dynamics reveals that the last kilometer of the delivery journey still predominantly adheres to the traditional door-to-door model. Couriers, who play a pivotal role in the process, contend with the immense daily delivery volumes. When compounded by the additional challenges posed by merchant promotions and seasonal spikes, the ability to deliver packages to residents within a limited time frame becomes increasingly constrained.

This scenario inevitably translates into a high demand for labor, substantial consumption of material resources, and time constraints that collectively affect the overall efficiency of the delivery process. Couriers often find themselves grappling with the pressure to fulfill delivery commitments while navigating the complexities of urban communities.

Moreover, the ongoing COVID-19 pandemic has introduced a unique set of challenges to the delivery landscape. Many communities have implemented stringent restrictions, prohibiting couriers from entering residential areas for door-to-door delivery. The primary objective of these measures is to reduce close contact and minimize the risk of transmission and infection. Consequently, express packages are often left unattended at the community's entrance or a designated station, awaiting collection by their owners.

However, this arrangement, while essential for public health and safety, is not without its shortcomings. Urban office workers, who form a significant portion of the population, are frequently engrossed in their professional commitments during the daytime, making it difficult to retrieve their parcels promptly. Additionally, elderly residents, particularly those living in high-rise buildings, may encounter mobility challenges that render descending to retrieve packages inconvenient or even impractical. This predicament further elevates the risk of packages going uncollected, potentially resulting in loss or damage.

In light of the profound impact of express delivery on our lives and the multifaceted challenges it faces, it is imperative to explore innovative solutions that can enhance the efficiency and convenience of last-kilometer delivery. As we navigate changing community dynamics and adapt to the ongoing impact of the pandemic, addressing these challenges becomes not only a priority but also an opportunity to ensure that express delivery remains a reliable and accessible service for individuals across all walks of life.

In this context, the future of express delivery is ripe for innovation, with potential solutions ranging from advanced automation and smart lockers to enhanced scheduling algorithms and tailored services. The journey of reimagining the last kilometer of delivery is an exciting and evolving one, as we work collectively to shape a future where convenience, efficiency, and accessibility define the express delivery landscape.

10.1.2 Delivery Problems Between Home and Community Pick-Up Point

The challenges surrounding last-kilometer express delivery in communities have reached a critical juncture, demanding innovative solutions that can address these pressing issues effectively. One such solution, proposed by smart logistics, is the development of unmanned trolleys—intelligent devices equipped with a range of advanced functionalities such as environmental recognition, autonomous decision-making, and active driving. This groundbreaking concept spans multiple technical fields, including mechanical structure, information technology, artificial intelligence, and more.

The unmanned trolley represents a significant leap forward in the realm of autonomous logistics. By adhering to preset modes and leveraging its autonomous capabilities, this smart device can execute specific actions within designated environments. It boasts a host of advantages, including automatic navigation, obstacle avoidance capabilities, robust stability, real-time responsiveness, and much more. These features position the unmanned trolley as an ideal candidate for revolutionizing the logistics and express delivery landscape.

Imagine a scenario where a courier vehicle, in the form of an unmanned trolley, can automatically transport parcels to designated households within a community, precisely on schedule and in alignment with the express delivery demand. This innovative approach not only streamlines the delivery process but also offers several invaluable benefits to both residents and service providers.

First and foremost, this automated system ensures that community residents receive their parcels promptly and reliably. With the ability to autonomously navigate and avoid obstacles, the unmanned trolley can navigate complex community layouts and reach even the most remote households, eliminating the common challenges associated with door-to-door delivery.

Furthermore, this solution enhances the overall efficiency of express delivery services, optimizing the allocation of resources and personnel. Couriers can focus on overseeing the unmanned trolley fleet, ensuring smooth operations, and addressing any exceptional situations that may arise.

Importantly, the adoption of unmanned trolleys aligns seamlessly with the imperative for epidemic prevention and control. By reducing the need for close contact between couriers and residents, it contributes to a safer delivery environment during times of heightened health concerns. This proactive approach aligns with public health guidelines and serves to protect both couriers and residents.

Ultimately, the introduction of unmanned trolleys into the world of express delivery promises to deliver unparalleled convenience and efficiency to the lives of community residents. By combining cutting-edge technology with the logistics industry, this innovative solution has the potential to redefine the last kilometer of delivery and shape a future where express delivery services are more accessible, reliable, and responsive than ever before.

10.2 Issues and Solutions of Automatic Community Delivery

By employing an extensive literature investigation approach, we have meticulously scoured a plethora of relevant articles and information, meticulously amassing an invaluable treasure trove of reference materials. This painstaking process has enabled us to delve deep into the intricacies of the express car phenomenon within the community context.

Venturing beyond the confines of desk-bound research, we embarked on immersive field investigations to glean firsthand insights into the characteristics and utilization prerequisites of these ubiquitous community express vehicles. We traversed the bustling streets and tranquil alleys, conducting surveys and interviews to unravel the multifaceted tapestry of this indispensable mode of transport.

Having thus amassed a wealth of empirical data, we harnessed the power of the comparative analysis method to scrutinize with meticulous detail the pros and cons of both the conventional vehicles currently in use and the revolutionary smart cars that we have meticulously designed. This analytical odyssey has facilitated a comprehensive understanding of the strengths and weaknesses inherent in these vehicular paradigms.

Incessantly driven by an unwavering commitment to refinement, we embarked on a relentless journey of iterative improvement, continuously refining and enhancing our smart car designs to align them with the evolving demands of the community's express delivery ecosystem. This unwavering dedication to perfection has yielded a product that stands at the forefront of technological innovation.

Finally, emerging from this exhaustive research and development process, we proudly present our groundbreaking vision and design blueprint, which strategically deploys smart cars within the realm of community express delivery. This pioneering approach is poised to revolutionize the way express deliveries are executed within the community, ushering in an era of unparalleled efficiency and convenience.

10.2.1 Implementation Obstacles of Automatic Delivery

Intelligent express car is mainly used for express logistics and service industry (hotel) automatic delivery. The following table analyzes and considers the advantages and disadvantages of express delivery vehicles (Table 10.1).

According to the above investigation, it has shown that there have been some applications of smart cars in shopping malls and hotels, but there are few applications in the community. There are mainly three types of problems.

Precise Navigation and Obstacle Avoidance

How do smart cars navigate to accurate buildings due to the complex community environment? The current accuracy of the commonly used navigation software, such as the Baidu map, can navigate to the community's entrance. Still, the specific building in the community on the map is seldom identified on the map. Therefore, the first problem that must be solved to achieve accurate distribution within the community is to mark the details of the community's roads and buildings on the car's built-in map.

There are all kinds of buildings in the district. Some doors need to be authorized. Some elevators are old and without a local area network. The problem of getting on and off the elevators needs to be solved.

User Interaction

Before delivery, it is necessary to confirm with the residents whether they are at home and whether it is convenient for the car to deliver the goods to their homes. It is also necessary to notify the customer to pick up the goods and inform them of the password for taking them to prevent the incorrect goods from being taken and to confirm that the goods have been received.

Table 10.1 Analysis and comparison of the current situation of smart car

Name	Main functions	Advantages	Disadvantages	Thinking
JD.com 4.0 Generation Intelligent Express vehicle	It has the capability of autonomous driving, intelligent obstacle avoidance, and traffic light recognition without manual intervention	The automatic driving system can sense the surrounding road environment and control the steering and speed of the vehicle according to the sensory information to drive safely and reliably and reach a predetermined place	They are mainly used for roads outside the community	If applied to the community, miniaturization improvement is required
Hotel automatic delivery robot	Automatic delivery to customers in the hotel building	It can be used well in the hotel's local area network	It is small in volume and capacity and can only be contacted by personnel through the local area network of the hotel	Need to extend the application scenarios outside the local area network
Community delivery robot	It can deliver goods in the residential yard	It can meet the requirement of zero contact in epidemic prevention	It cannot go upstairs, and residents must go downstairs to pick up the express	Need to resolve in-building delivery issues
Robots with moving and operating mechanical arms	Use the mechanical arm to solve the problem of pressing the elevator	Robots can get on and off the elevator	Low load capacity and slow response of the robotic arms to press the elevator	Smart cards can be used in place of robotic arms and combined with a dispensing bin

10.2.2 Core Features of the Community Smart Car

Based on the above problems, and some data and information are studied and analyzed, the following optimal design ideas are put forward for the intelligent express car in the community.

10.2.2.1 Route Planning and Obstacle Avoidance

To realize automatic delivery, it is necessary to solve the problem of route planning and optimization, mainly solved by the navigation module in the smart car. When placing articles into the intelligent car cabinet, the delivery personnel shall scan the QR code through the car, and the car will automatically open an empty cabinet. The delivery personnel shall put in the goods and close the cabinet. When the delivery man puts in all the goods or the car cabinet is wholly occupied, the car automatically calculates and optimizes all paths according to the user's exact address obtained by the two-dimensional code and determines the priority of the delivery to the customer. The main algorithm principle is that the time for delivering goods is minimal. When the sequence and the route are planned, the smart car can navigate according to how to reach the destination safely. I learned the method of making a navigation map through studying. I can shoot all the community's roads, buildings, and corridors with a camera to create a high-definition navigation map of the community, mark the household number of the resident on the map, and then build it into the intelligent car navigation module to support offline navigation. In addition, the building and floor of the smart community are equipped with Wi-Fi equipment, and the smart car can carry out accurate positioning and navigation in the building with the Wi-Fi positioning technology.

From the starting position to the designated place, the intelligent car must be able to bypass the obstacle on the travel route during the moving process, which can be solved through the sensing module. By studying relevant literature, we have learned that the existing robot operating system (ROS) can realize the perception and collection of environmental information through lidar and depth cameras and then process the acquired environmental information, which can be quickly identified after encountering obstacles, and form a safe and fast new circuit diagram which can avoid obstacles.

10.2.2.2 Access Control and Elevator Access

This intricate challenge necessitates a dedicated focus within the realm of autonomous driving technology, finding its solution only within specific, localized environments seamlessly integrated into a local area network. In the contemporary landscape of residential developments, the emergence of cutting-edge housing estates equipped with state-of-the-art intelligent access control systems has opened up a promising avenue for innovation. Within this context, the smart car emerges as a pivotal constituent of the community's intelligent mobility infrastructure, poised to revolutionize the way we envision urban living.

A visionary approach to surmounting this challenge entails the seamless integration of the smart car's communication module with the existing intelligent access control system, thus forging an interconnected ecosystem. This groundbreaking solution envisions a scenario where, as the smart car arrives at the threshold of a building, the intelligent access control system, attuned to its presence, engages inductively, culminating in the automatic unlocking of the access point. This streamlined interaction not only enhances the efficiency of access but also underscores the synergy between cutting-edge transportation and intelligent residential infrastructure.

Moreover, the smart car, akin to a technological virtuoso, extends its capabilities to the very heart of the building itself. With its integrated smart card system, the vehicle becomes a conduit for information exchange with the building's elevator network. This transformative feature enables the seamless navigation of the vertical expanse of the building, affording residents a level of convenience hitherto unattainable. The smart car's harmonious integration with the building's infrastructure ushers in an era where mobility is not just confined to the streets but extends effortlessly to the domains of urban living.

In sum, the challenge at hand finds its resolution within the realm of autonomous driving technology, where the smart car emerges as a catalyst for a harmonious interplay between modern housing estates and cutting-edge transportation. Through the integration of communication modules and intelligent access control systems, the smart car paves the way for a future where urban living is seamlessly intertwined with intelligent mobility, redefining the very fabric of our communities.

10.2.2.3 Item Access

Incorporating cutting-edge technology into the logistics process, our smart car has been equipped with intelligent barcode scanning capabilities, offering a transformative solution to the perennial storage challenge. When a courier deposits an item into the smart car's cargo compartment, the vehicle springs into action. Employing its sophisticated barcode scanning technology, the smart car promptly identifies the pertinent information associated with the item, including details about the object itself and the contact telephone number of the owner. This seamless process ensures that each item is meticulously cataloged and allocated to its designated storage cabinet, thereby resolving the storage conundrum with unrivaled precision.

Upon reaching the doorstep of the intended recipient, the smart car's information communication module takes center stage. Leveraging its connectivity prowess, the module initiates contact with the item's owner, facilitating seamless communication. Through this interaction, the smart car imparts a unique password to the recipient and a digital key to unlock the cargo compartment securely housed within the vehicle.

With this password in hand, the item's owner interfaces with the smart car's intuitive touchscreen interface, entering the confirmation code provided. This act triggers a meticulously orchestrated sequence of events within the vehicle's cargo compartment. The corresponding storage cabinet gracefully unlocks, granting access to the awaiting item. Once the item has been retrieved, the cabinet door gracefully swings shut, securing the remaining items within.

In instances where the recipient is absent or has been notified of a delay in pick-up, the smart car assumes the role of a temporary custodian. Safeguarding the items within its secure confines, it patiently awaits the next opportunity for delivery.

Upon successfully completing this intricate process, the smart car's onboard navigation system springs back into action. It diligently re-optimizes its path, ensuring that the subsequent delivery is executed with maximum efficiency. This continuous cycle of intelligent, automated logistics exemplifies the future of streamlined, technology-driven delivery services, where convenience and precision converge to redefine the very essence of courier operations.

10.2.2.4 Environment-Friendly Effort

Our unwavering commitment to environmental protection and sustainable energy solutions remains a paramount concern in our smart car innovation. To address these pressing issues, we are fervently exploring alternative energy sources to reduce our reliance on traditional fuels. While the primary energy source for our smart cars remains electricity, we have introduced an ingenious solution to harness the power of the sun.

In our quest for a greener future, we have integrated solar panels onto one side of the smart car's body. These solar panels serve a dual purpose: they actively charge the vehicle while it's in motion and automatically disengage when not needed. When the smart car is at rest and its electric charge stands at or above 50%, it seamlessly transitions to solar charging, harnessing the sun's energy to replenish its power reserves. This groundbreaking innovation not only conserves energy and mitigates environmental impact but also significantly reduces the risk of the vehicle running out of power during operation.

Further underscoring our environmental commitment, our smart cars employ an offline navigation mode within the community. This innovative approach minimizes reliance on wireless networks, thereby reducing power consumption and mitigating signal radiation. By opting for this energy-efficient navigation mode, we are contributing to both a more sustainable operation and a healthier urban environment.

Moreover, our dedication to eco-conscious design extends to the materials used in crafting the smart car's components. Whenever possible, we prioritize the use of environmentally friendly materials, ensuring that our vehicles leave a minimal ecological footprint. Additionally, we prioritize waterproofing for these components, enhancing their longevity and resilience in various weather conditions, while also reducing the need for frequent replacements and thereby promoting sustainability.

In essence, our smart cars stand as a testament to our unwavering commitment to environmentally responsible innovation. By harnessing solar power, optimizing energy efficiency, and employing eco-friendly materials, we are paving the way for a more sustainable and eco-conscious future in the realm of smart mobility.

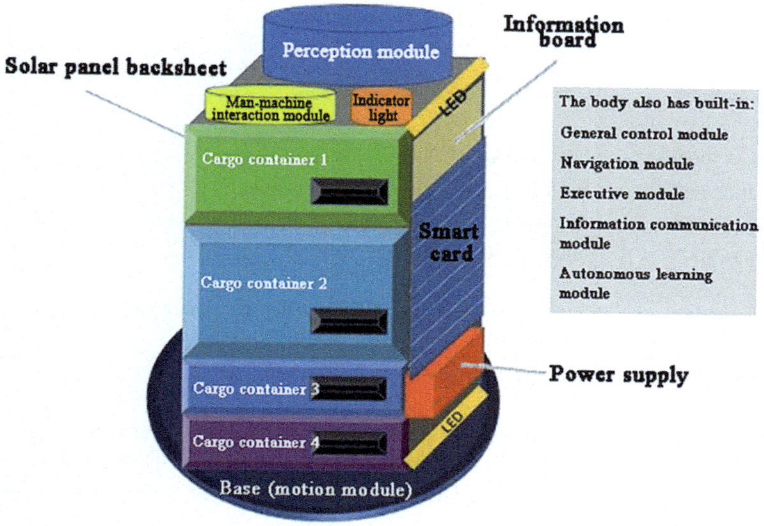

Fig. 10.1 Schematic diagram of the smart express car

10.3 Design and Simulation of Smart Car Prototype

10.3.1 Function Design

Based on the above ideas, we have designed the Smart Express Car, which mainly includes the general control module, the execution module, the navigation module, the information communication module, the sense module, the man–machine interaction module, and the autonomous learning module. The schematic diagram and function are described as follows (Fig. 10.1 and Table 10.2).

10.3.2 Operational Simulation Design

Once an express delivery item is securely placed inside the Smart Express Car's designated cabinet, a unique pickup code is promptly generated, setting the stage for an efficient and secure delivery process. With the cargo loaded and the mission clear, the Smart Express Car springs into action.

First and foremost, it initiates the process of route planning and optimization. Leveraging advanced algorithms and real-time data, the vehicle meticulously calculates the most efficient path to navigate through the community, taking into account factors like traffic conditions and delivery time constraints. This intelligent route planning ensures timely and effective delivery operations.

Table 10.2 Main functional modules of the smart express car

Module	Function
General control module	Central information processor
Executive module	Send command to each module
Navigation module	Built-in accurate maps for path planning
Information communication module	Realize the information transmission to the outside world and interaction with the control center
Sense module	Sensing the change in the external environment and providing real-time information
Human–machine interaction module	Communicating with people around
Autonomous learning module	With the help of artificial intelligence, improve the ability to receive and analyze the data so that the car can react more quickly and reasonably
Built-in ultraviolet disinfection lamp	The ultraviolet disinfection lamp is placed at the top of the inside of each goods cabinet. When goods are detected, automatic disinfection is carried out to ensure the cabinet door is closed

As part of its preparatory checks, the Smart Express Car also conducts an automated assessment of its electric power reserves. By estimating the total route length and anticipated delivery time, the vehicle gauges whether its current electric charge will suffice for the entire mission. If the projection indicates an electric reserve of no less than 30% after completing the deliveries, the car proceeds to embark on its journey from the starting point.

For instance, consider a scenario where the Smart Express Car needs to deliver items to residential building A. The vehicle meticulously follows its optimized route plan, selecting the most efficient path available, such as Route A, as determined by its algorithms. As it navigates this path, the car utilizes advanced technologies like laser radar and depth cameras for real-time obstacle detection and avoidance. This allows the vehicle to gracefully sidestep pedestrians, vehicles, and objects that may unexpectedly obstruct its route.

In the rare instance where the vehicle, objects, or pedestrians pose an insurmountable obstacle within the designated time frame, the Smart Express Car's adaptive intelligence comes to the fore. It swiftly reconfigures its route, autonomously planning an alternate path to circumvent the impediment, ensuring the safety and efficiency of its journey.

In essence, our Smart Express Car is a testament to the fusion of cutting-edge technology and adaptive intelligence. It is engineered to navigate through the dynamic complexities of a community environment, ensuring reliable and efficient express deliveries while prioritizing safety and optimal energy utilization (Fig. 10.2).

Once the smart car gracefully enters the confines of a building's entrance, its remarkable communication module, exemplified by the smart card, springs into action, orchestrating a seamlessly choreographed ballet of precision and efficiency.

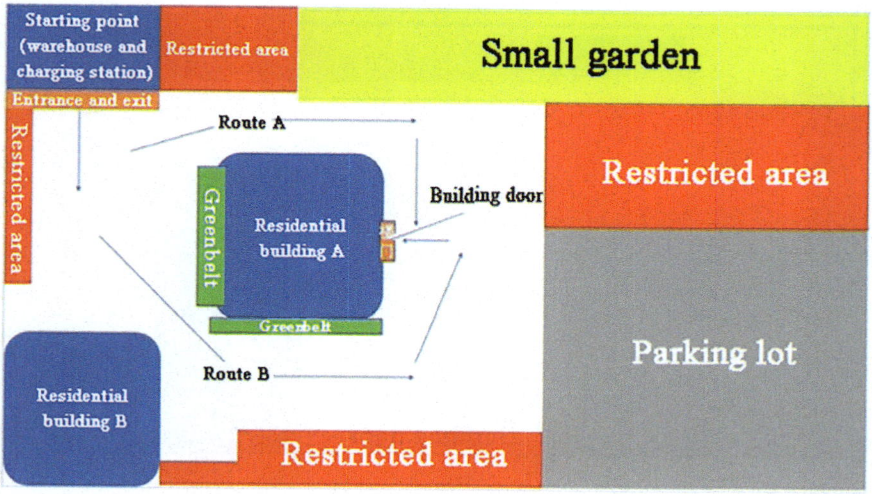

Fig. 10.2 Schematic diagram of route planning

The communication module initiates contact with the building's elevator system, establishing a digital dialogue that transcends conventional expectations. With the innate ability to analyze the real-time movement of multiple elevators within the building, it astutely determines which elevator is poised to reach the first floor most expeditiously. With this knowledge in hand, the smart car autonomously proceeds to the designated elevator, positioning itself with precision and poise.

As the smart car and the elevator converge, the communication module continues to interact with the elevator system, sharing essential information. This includes transmitting the desired floor information to the elevator, effectively synchronizing the vehicle's journey with the elevator's ascent.

With synchronized precision, the elevator effortlessly transports the smart car to the specified floor, seamlessly coordinating the opening of its doors with the car's arrival. As the elevator doors elegantly part, the smart car effortlessly glides out, poised to embark on the final leg of its journey.

Guided by an indoor navigation route, skillfully maneuvered through a fusion of the navigation and information communication modules, the smart car navigates the interior of the building with grace and precision. It traverses corridors, hallways, and pathways with the same level of accuracy and intelligence it demonstrated on the community streets outside.

This harmonious symphony of technology, featuring the smart car's navigation prowess and the communication module's seamless interaction with elevators, ensures that express deliveries reach the doorstep of residents within the building, embodying a seamless fusion of mobility, technology, and efficiency (Fig. 10.3).

Navigating to the third floor of the building with unwavering precision, the Smart Express Car leverages the floor navigation map and the meticulously optimized route to arrive at the user's doorstep. This advanced operation unfolds with the seamless

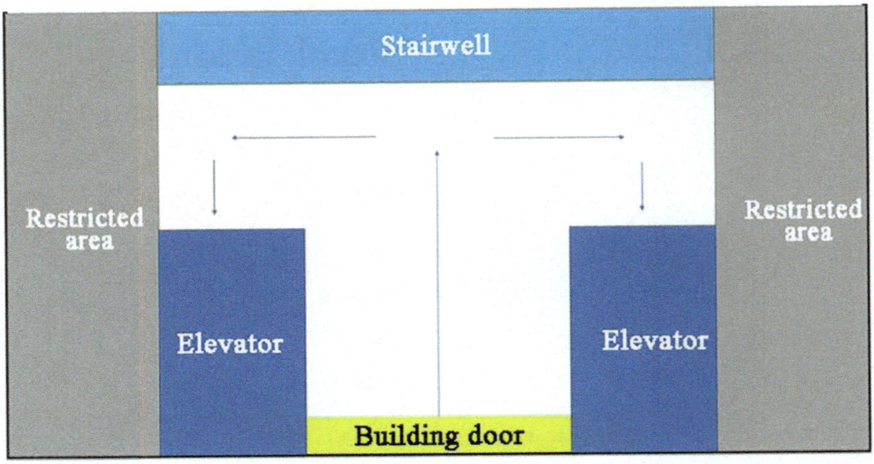

Fig. 10.3 Plan of the first floor

coordination of the information communication and human–computer interaction modules, ensuring a streamlined and user-friendly experience.

Once the Smart Express Car reaches its destination, it initiates contact with the intended recipient. Through the information communication module, the vehicle sends a notification to the user, informing them that their delivery has arrived and is ready for pick-up. In this message, a unique picking code is provided, serving as the digital key to unlock the cargo compartment securely housed within the vehicle.

The user, receiving the notification, proceeds to the Smart Express Car to collect their goods. With the provided picking code, they interface with the human–computer interaction module, entering the code on the car's intuitive touchscreen interface. This simple and secure process triggers the unlocking of the designated cabinet, granting access to the eagerly awaited items.

However, in the event that the user does not pick up the goods within a specified time frame, set at a generous 3-min window, the Smart Express Car exhibits its adaptability and efficiency once more. Recognizing the unattended delivery, the vehicle takes the initiative to return to its original route, ensuring that deliveries continue without undue delay.

Simultaneously, the information communication module sends a gentle reminder to the user, conveying information about the next available pick-up time. This proactive communication ensures that users are well-informed and can conveniently schedule their pick-up in alignment with their daily routines.

In essence, this elegant orchestration of technology not only streamlines the delivery process but also ensures that users have the flexibility and convenience they need to retrieve their goods at their own pace. The Smart Express Car, driven by the information communication and human–computer interaction modules, exemplifies the pinnacle of efficiency, adaptability, and user-centric design in the realm of express delivery (Fig. 10.4).

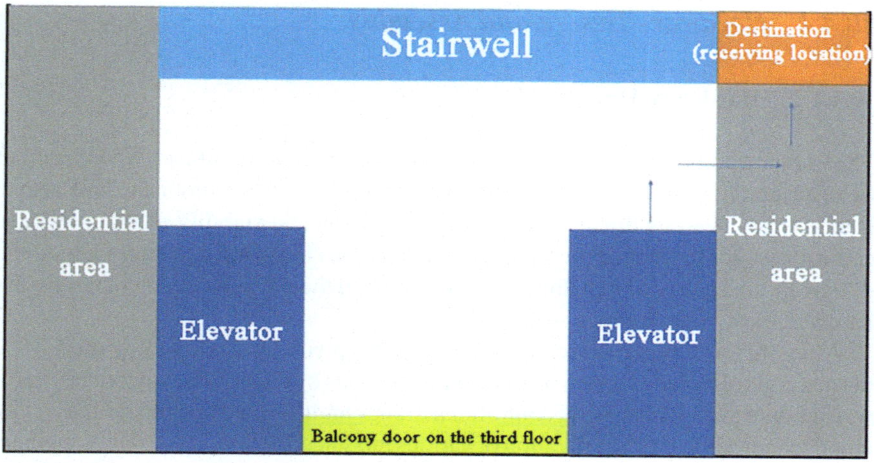

Fig. 10.4 Plan of the third floor

10.4 Implementation of Smart Car Prototype

Based on the above research, we assembled an intelligent car under the teacher's guidance and debugged and ran it, and it realized most of the functions in our conception. The Smart Express Car integrates a plurality of technologies. There is a general control module, which is the car's general control system. As the center of the control, with the help of a computer system, all the units need to integrate and interact with the available control module. The controller processes the data sent by the sensor module and then sends it to the drive module, thus controlling the movement of the whole car and finally completing the automatic operation and control process. This trolley mainly comprises a control unit, an execution unit, a navigation module, a driving unit, a road identification and avoidance module, an information communication module, a human–computer interaction module, etc. The flow chart of the smart car's control system is shown in Fig. 10.5.

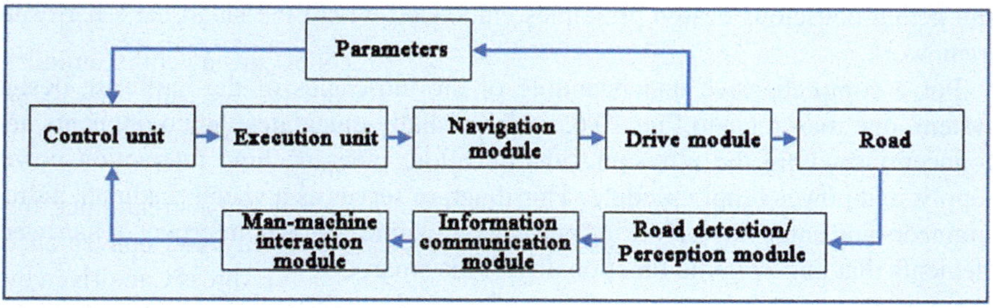

Fig. 10.5 Flow chart of control system

10.4.1 Hardware Design and Assembly

10.4.1.1 Hardware Design

The implementation of hardware design holds paramount importance when it comes to ensuring the seamless and efficient operation of smart automobiles. This significance arises from its direct impact on the precision and stability of the integrated control system. In light of this criticality, it becomes imperative to adopt a modular design approach, a paradigm that can be considered the cornerstone of contemporary hardware design principles.

At the heart of this research endeavor lies the pivotal role played by the ESP32 master control board, serving as the central nervous system of the smart car's hardware architecture. This central controller assumes multifaceted responsibilities, chief among them being the orchestration of various modules, facilitating seamless connectivity within the Internet of Things (IoT) ecosystem, interlinking disparate hardware modules, and overseeing image acquisition through the dedicated image acquisition module, colloquially referred to as the video board.

The image acquisition module, or video board, is responsible for the sophisticated task of capturing visual data, thus empowering the smart car to identify objects in its vicinity. This identification process culminates in the issuance of precise commands to the central control panel, a core element in the smart car's decision-making process. It is the central control panel that acts as the conductor, directing the vehicle's responses and actions based on the visual information received.

Intriguingly, hardware design extends its influence to the realm of physical maneuverability as well. The steering engine driving module assumes the critical duty of propelling the smart car, enabling it to navigate and traverse its environment with precision and grace. Furthermore, this module also takes charge of managing the car's essential functions, such as opening and closing the cabinet door, thus showcasing the versatility of hardware in this intricate system.

Once the cabinet door gracefully closes, a strategic and hygiene-focused feature comes into play. It is the ultraviolet switch, a pivotal component within the hardware architecture. Upon the cabinet door's closure, this switch is seamlessly activated, triggering a robust disinfection process. This safety measure underscores the intelligent and health-conscious design principles embedded within the smart car's hardware framework.

For a comprehensive understanding of the intricacies of the hardware design system, one may refer to Fig. 10.6, which visually elucidates the components and connections within the ultrasonic identification, man–machine interaction power supply, and power supply module. This diagram serves as a visual roadmap, aiding engineers and enthusiasts alike in comprehending the complex interplay of hardware elements that underpin the functionality of the smart car.

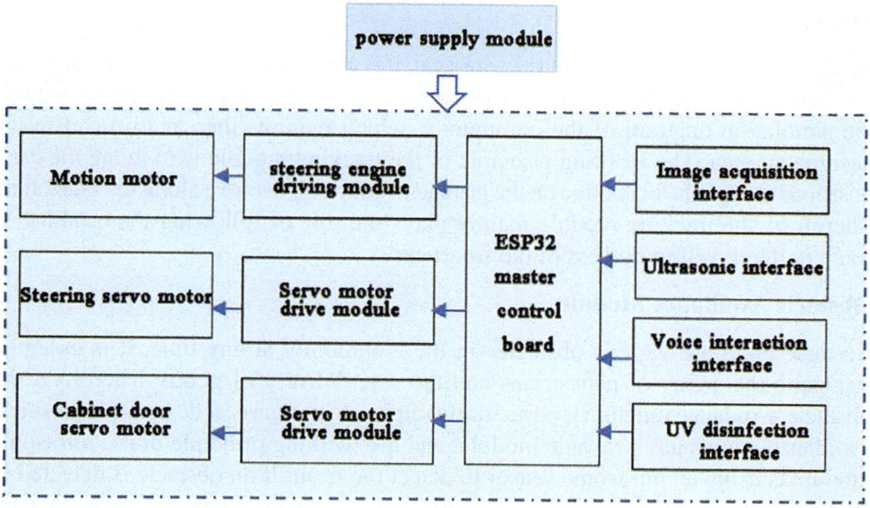

Fig. 10.6 Hardware design system diagram

Electric Drive

First, the design of the power circuit is the power source of the smart cart, providing power for the car's operation. Based on the characteristics of the ESP32 controller, the whole car can be operated only by supplying power to ensure its stable working voltage. In the preliminary debugging and testing process, the USB port can be connected to the host computer to work normally and facilitate the inspection. The portable charger can be used when the assembly is completed and the access program is debugged and calibrated. The power module of the car uses four No. 5 batteries and a portable charger as the power supply. It uses four DC motors to realize the mutual conversion of DC electric energy and mechanical energy. For the smooth running of the car, four wheels are selected in this design.

The second is the steering engine interface circuit. In the design of the smart car, the steering gear is mainly used to ensure the vehicle's steering. Therefore, the operating voltage, steering action, and steering speed of the steering gear are all factors to be considered.

The smart car mainly adopts four motors to drive four wheels respectively, and the car operation has a left turn, right turn, straight line, and stop four states. It uses the tracking module to detect the road signal and transmit it to the general control module for real-time control. It outputs the corresponding signal to drive the DC motor and runs by adjusting the speed of both motors. In the running process, if the ultrasonic module detects an obstacle, the signal is also transmitted to the master control module. The master control module will send a stop command to stop the car.

Tracking Module

Based on the design function of the smart car, it is also necessary to simulate the route tracking of the community. According to the route of the community, we designed the simulation diagram of the community, which requires the car to travel in the designated area. The working principle of the tracking module is to judge the car's position through the black line on the ground to control it to move along the black line. Therefore, the tracking module mainly plays the role of following the established route in the traveling process of the smart car.

Obstacle Avoidance Module

Because there are various obstacles in the community at any time, it is essential to ensure the safety of pedestrians and the safe delivery of goods. Therefore, the obstacle avoidance function is particularly important in smart car design. The obstacle avoidance adopts an ultrasonic module, and the working principle of the ultrasonic module is using an ultrasonic sensor to detect the front. If an obstacle is detected in the front, the obstacle is avoided. The ultrasonic module is connected to the general control module by an ultrasonic sensor, and the main function of the ultrasonic sensor is to measure the distance.

Communication Module

After the smart car delivers the goods to the home, it is required to interact with the user to send the pick-up password to the user. The corresponding cabinet can be opened only when the correct pick-up password is input. In addition, when the goods are stored in the cabinet, the sorter can also use the voice module to carry out dialogue, for example, to access large, medium, or small articles, and the corresponding cabinet door is automatically opened through voice interaction. Therefore, the intelligent car must be internally provided with a module for transmitting and receiving information like a mobile phone.

In the smart car's design, a voice sensor module is used for voice recognition. The working principle of the module is mainly to use the interconnection of Wi-Fi and a local area network. After the receiver's mobile phone number is configured in the program, voice recognition and interaction are completed through the voice interaction module. The voice recognition module adopted is simple and does not need to be trained and recorded in advance, and the recognition rate can reach more than 95%. In addition, it also uses the Blinker Lighting App to simulate the software and connect the user's mobile phones. This app is a mobile phone access tool specially designed for the Internet of Things, which can make the users more convenient to operate the hardware equipment.

10.4.1.2 Smart Express Car Assembly

To produce a car model which is consistent with the design function and can run, we use a 3D printer to print a car body assembly board and assemble four steering

Fig. 10.7 Actual effect picture of smart car

engines and wheels, four container steering engines, an ESP32 general control board, a voice module, an intelligent video recognition integrated module, two ultrasonic ranging modules, and a tracking module. In addition, according to the actual situation of the community, a simulation map of the community is designed, and the plane scene diagram of the community is printed. Then correctly connect each module.

The actual effect of the smart car assembly is shown in Fig. 10.7.

The main difficulties and solutions encountered during assembly are as follows:

Wood board is easy to deform

The primary assembly material is the 3D-printed wood board during assembly, so it is easy to deform slightly. It led to a bit of depression in the baffle of the bottom cabinet door, which affects the installation and rotation of the bottom cabinet door, making it easy to jam. The final solution is to install the lower cabinet door first, but the installation position can be staggered a little to prevent the door from clogging and damaging the motor. Each of the other three cabinet doors is provided with a module motor. Even if one cabinet door cannot be used, the overall test effect will not be affected.

The smart car is loaded too heavily

Because the car needs to load goods and its portable charger, if the counterweight of all kinds of integrated modules is too heavy, it will lead to problems of speed deviation, poor steering, and no start. The car was tested for motion using only one chassis, and the bodywork was not installed. After the car body is installed, the counterweight is significantly increased, which leads to the task that the car cannot run to the designated place and identify the road sign in the specified time due to the slow-down speed. Hence, the car turns when it should not and easily runs out of the selected area. The solution found later was to adjust the procedure to solve the start-up and steering problems through the speed allocation.

10.4.2 Function Implementation of Smart Car

The software design aspect of the smart car is an equally pivotal facet of its operational prowess. It revolves around the intricate world of programming and algorithm development, which are integral components within the car's sophisticated control system. Broadly speaking, the software design can be broken down into several key elements, including the main program, voice interaction program, and video recognition program. These components collectively serve as the brainpower behind the smart car's intelligent decision-making and actions.

First and foremost, the main program stands as the central orchestrator of the software architecture. It acts as the conductor, coordinating the various software modules and ensuring their seamless interaction with the hardware components. The main program's design is crucial, as it sets the foundation for the entire software ecosystem, enabling efficient communication between different software and hardware elements. This program is akin to the smart car's neural hub, governing its responses to external stimuli and internal commands.

In addition to the main program, the voice interaction program plays a pivotal role in enhancing the smart car's human–machine interaction capabilities. It empowers the car to understand and respond to spoken commands or inquiries, thus making the user experience more intuitive and user-friendly. The voice interaction program's design necessitates advanced speech recognition algorithms and natural language processing techniques, allowing the smart car to communicate effectively with its human passengers.

Moreover, the video recognition program is a cutting-edge component that bestows upon the smart car the ability to perceive and interpret its surroundings through visual data. This program employs sophisticated computer vision algorithms to analyze the information captured by the image acquisition module, which was earlier mentioned in the hardware design section. It enables the smart car to identify objects, obstacles, and relevant environmental cues, thereby facilitating safe navigation and informed decision-making.

A key hallmark of the software design process is the utilization of Mixly, a graphical modular programming software. Mixly's user-friendly interface and intuitive design philosophy simplify the coding process, making it accessible even to those without extensive programming expertise. This choice of software tool exemplifies a commitment to efficient and streamlined software development, allowing for the rapid creation, maintenance, and debugging of the smart car's control system.

In essence, the software design of the smart car represents the intricate synergy of programming and algorithmic ingenuity. It converges technology and innovation to create a control system that not only operates efficiently but also interacts intelligently with users and the surrounding environment. Through the careful design of the main program, voice interaction program, and video recognition program, the smart car embodies the forefront of cutting-edge automotive technology, ushering in a new era of transportation innovation.

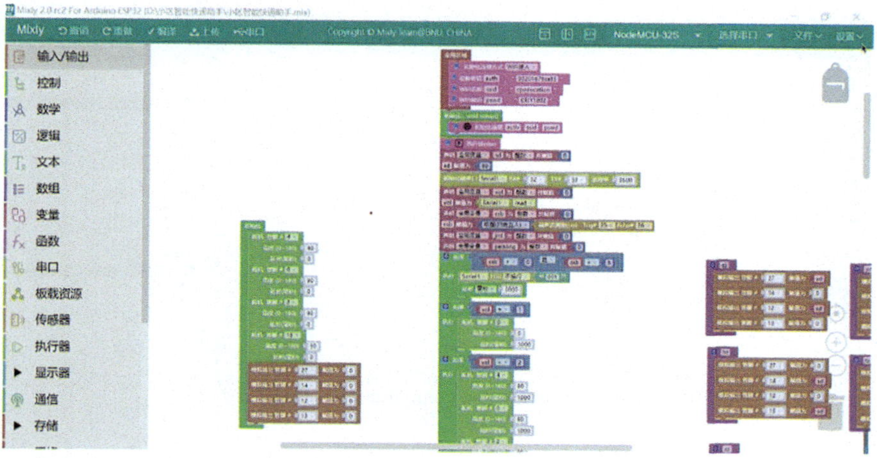

Fig. 10.8 Main programming-related programs and interfaces

10.4.2.1 Program Design

The main program is to design the overall software of the car. For the different modules, the main program must initiate the data parameters, image processing, path identification algorithm calculation, motor control algorithm calculation, steering engine control algorithm calculation, image data collection and processing, etc., when designed.

Configuring the Network

The primary function accomplished is to integrate and link the modules. First, complete the configuration of the local area network (LAN), the integration and linking of Blink App and different modules, integration, and video recognition and interaction.

Configure the Steering Gear

The primary function is to control the opening, closing, and steering of the container door by the steering engine under different conditions. The main procedure is the setting of variables and the configuration of the motor and steering gear. The related programs and interfaces related to the main programming are shown in Fig. 10.8.

10.4.2.2 Voice Interaction Program Design

Regarding the design of the voice interaction program, the primary function completed is to interact with users. The commonly used terms must be preset first in the voice recognition mode. When picking up or delivering the express, the user

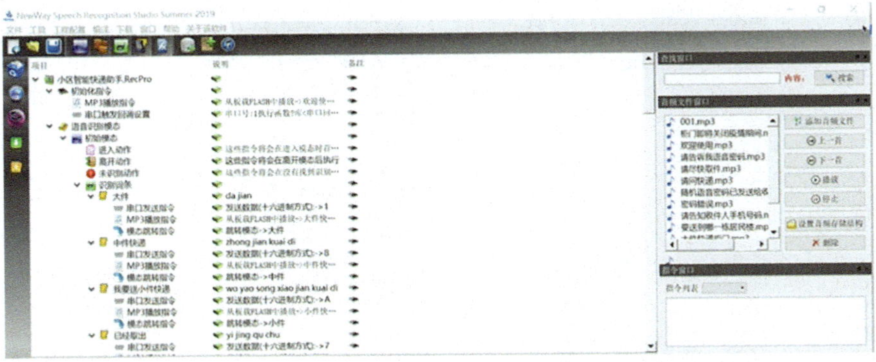

Fig. 10.9 Main entries for main program and configuration

speaks a similar command, and the voice module receives it and executes the follow-up program. The relevant terms must be preset in the software, which can be increased or decreased according to the scene. For example, 'Welcome to use the intelligent community express assistant' is played after the machine is turned on. There are some terms, such as 'Is it a large item or a small item?,' 'Please tell the pick-up password, 'The cabinet door is about to be closed,' etc. 12 terms are pre-configured in the speech recognition program of this smart car. The main procedures and entries are shown in Fig. 10.9.

10.4.2.3 Video Recognition Program

The video recognition program within the smart car's software architecture is tasked with a critical mission: the identification of key landmarks, including but not limited to residential buildings A and B, and the determination of the appropriate path to follow, whether it entails making a left turn or a right turn. This aspect of the software is a fundamental component, as it directly contributes to the smart car's ability to navigate its environment effectively.

Achieving a high degree of accuracy in image processing and recognition is paramount to the success of this program. To attain such precision, an extensive dataset comprising more than 800 images is employed for machine learning purposes. These images serve as the foundation upon which the recognition algorithms are trained. The process involves leveraging the built-in functionality of the video module, which facilitates the reading and utilization of this comprehensive image dataset for training.

Several critical factors influence the accuracy of video recognition, with the foremost being the quantity and diversity of training images. To enhance recognition capabilities, it is imperative to collect a substantial number of images, capturing various scenarios, light conditions, and angles. This diversity in the dataset equips the recognition program with the adaptability and robustness needed to perform effectively under real-world conditions, where lighting and perspectives may vary significantly.

The impact of different lighting conditions on recognition accuracy cannot be understated. Consequently, the smart car's image acquisition system is designed to capture images under a wide range of light rays, thereby exposing the recognition program to the various lighting scenarios it might encounter in its operational environment. This broad spectrum of lighting conditions is essential to ensure that the recognition algorithms can perform reliably, whether it's a bright, sunlit day or a dimly lit evening.

Furthermore, during the debugging and fine-tuning phase of image processing, the recognition program incorporates features to enhance accuracy. This includes techniques like framing the detected areas of interest within the images and labeling relevant information. Such visual cues, such as object type labels, serve to guide the program in its recognition efforts, enabling it to pinpoint and classify objects with greater precision.

In sum, the video recognition program is a cornerstone of the smart car's software design, enabling it to make informed decisions regarding its path and surroundings. To achieve the desired accuracy, the program relies on a substantial training dataset, diverse lighting conditions, and refined debugging techniques. This meticulous approach ensures that the smart car's image recognition capabilities are robust and reliable, paving the way for safe and efficient autonomous navigation.

10.4.2.4 Tracking Program Design

When the sensor on one side of the tracking module senses a black track, a signal 1 is transmitted to the main control board through the tracking module so that the car can steer in time at a higher speed without deviating from the track due to slow steering when encountering a large curve.

10.4.2.5 Design of Obstacle Avoidance Program

When the ultrasonic sensor detects an obstacle, a signal is transmitted to the main control board through the obstacle avoidance module to control the rotation direction of the DC motor. Considering that obstacles have different sizes and shapes, this design always stops the movement first when encountering obstacles. The flow of obstacle avoidance is shown in Fig. 10.10.

10.4.3 Debugging of the Smart Car

The debugging process and contents of the smart car include the following points: firstly, the mechanical performance of each module of the system shall be tested and

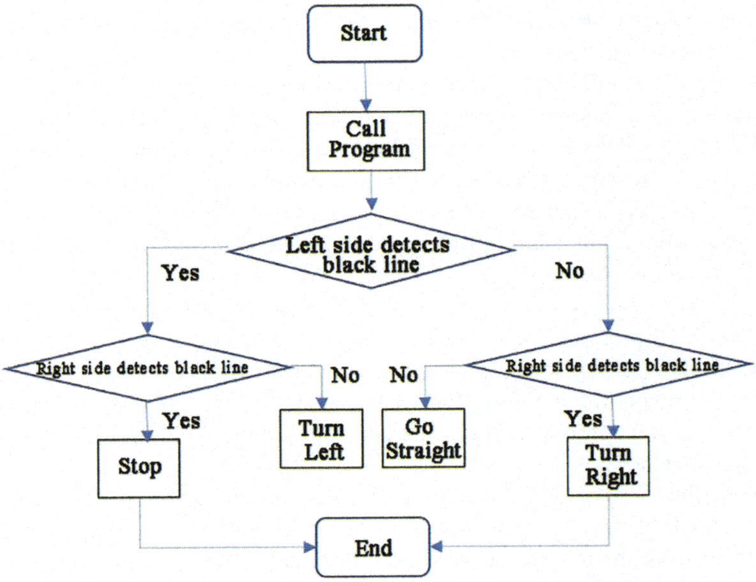

Fig. 10.10 Flow chart of obstacle avoidance

debugged, whether the connection between the equipment in each link is normal, and special attention shall be paid to whether the installation of the plug-in part is correct and dependable. Secondly, debug the program, upload the written program to the main control board, and debug whether each code in the system normally operates and whether each module can run normally. Moreover, finally, debugging the whole machine, aiming at adjusting the parameters of each module to meet the requirements, verifying voice recognition, route identification, obstacle identification, curve strategy program debugging, etc., in the process, continuously adjusting and optimizing the parameters in the running process, and finally completing the task of carrying goods by the intelligent trolley.

10.4.3.1 Tracking Debugging

At first, it is considered that the tracking module can execute the corresponding program after the black line is identified when the schematic diagram is made. However, after subsequent debugging, it is found that when the speed of the smart car is high or the angle is too small, the tracking module is not easy to identify the black line on the ground. The situation in which the smart car runs out of the planned route may occur. After analyzing the cause of the problem, we believe that if the hardware cannot be changed, the accuracy of tracking module identification can be improved by boldly planning the black line at the edge of the route.

10.4.3.2 Power Debugging

Although the tracking module was available, its capability could not enable the car to identify the ground signs faster to make a parking and steering action. Later, to solve this problem, we debugged the program for a long time and finally found a better speed to normally make the car run under the large counterweight. However, because the solution to this problem was to debug a program, there would be some derivative problems, and one of the more significant problems was that the car did not start fast enough. To avoid the problems caused by the adjustment of the procedure later, a set of procedures was added to set the starting speed of the car.

10.4.3.3 Video Training and Debugging

A video module cannot directly realize the function of identifying road signs and needs to give it some data in advance to train by itself. I took more than 800 photos of each road sign in different light and shade conditions and tested its accuracy after training, and the result was satisfactory. Since the pictures for training were taken in natural light in the daytime, the recognition rate was not exceedingly high at night under the light, but it did not affect the final use.

10.4.3.4 Production and Modification of Simulation Diagram

Although the original version of the simulation diagram design was derived from a real community, it was not an optimal choice for the environment of the car simulation. Therefore, a more straightforward plot plan was selected for simulation, as shown in Fig. 10.11.

The rectangular driving route is beneficial to the program design and path optimization of the smart car, reduces a certain number of unnecessary programs while also improving the convenience of simulation, and can see various situations in the running process of the smart car more intuitively. Moreover, if a problem is found, it can be debugged clearly and conveniently after it is solved.

10.4.3.5 Debugging Result

Through continuously optimizing the program, the smart car finally realizes the effects: It identifies the large, medium, and small-sized goods to be loaded through voice conversation from the starting point, opens the corresponding cabinet door is opened, reminds the delivery man to put in the goods as soon as possible and close the door, and automatically turns on an ultraviolet disinfection lamp after the door is closed; it sends a random voice password to a receiver after inquiring the residential building the goods to be delivered and the contact information of the consignee; the receiver clicks the start delivery on a mobile phone to start the delivery, and

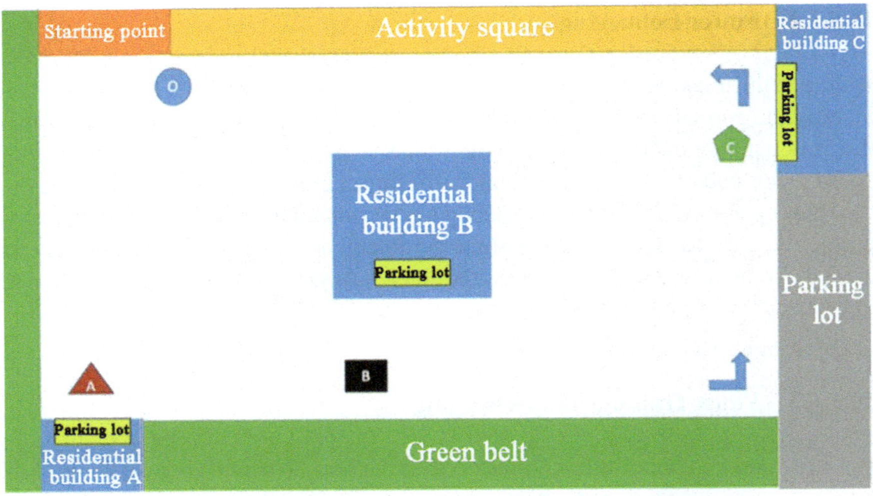

Fig. 10.11 Plot plane simulation diagram

the smart car automatically stops moving if obstacles are encountered on the road; the receiver will be inquired the voice password of the recipient after arriving at the destination, and if the password is correct, the corresponding cabinet door is opened, and the recipient is reminded to take out the goods as soon as possible and do hand disinfection; after the cabinet door is closed, the ultraviolet disinfection lamp is turned on again and returns to the starting point. The task of transporting the goods is completed (Fig. 10.12).

Fig. 10.12 Smart car production and debugging process

10.5 Conclusion and Future Plans

The application value of the smart car has been demonstrated in some fields. This research uses literature inquiry, field investigation, and comparative analysis to analyze the application demands and feasibility of the smart car in community express delivery. The application development research is carried out according to the smart car's technical characteristics and operation scheme. According to the relevant technologies, the smart car is innovated and improved. The functions of information interaction, path planning, automatic obstacle avoidance, self-learning, ultraviolet disinfection, etc., are mainly realized, which can meet the needs of residents in the community to pick up the express, facilitate the courier, and the requirements of epidemic prevention. In practice, the smart car also needs to be perfected and improved. It will enhance the express mode to a certain extent, reduce the cost of workforce, material resources, and time, improve express efficiency, and have good social and economic value. From the investigation of the project to the determination of the research scheme and then to the design and operation of the smart car, we have not only devoted to the project, expecting that the research can serve the practice, but also followed the clear research idea, adopted the appropriate research method, and solved the problem with the relevant knowledge of the smart car. In the research process, we are also constantly growing, gradually improving our essential character and critical ability to meet the needs of personal and social development.

Bibliography

Chen, S. (2019). *Research on indoor mobile robot map construction and path planning based on ROS*. Master's thesis, Guangdong University of Technology. https://doi.org/10.27029/d.cnki.ggdgu.2019.000497

Luo, D., Zhao, Z., & Yue, Y. (2017). Development status and trend of intelligent small cars. *Henan Science and Technology, 23*, 92–93.

Ren, F. (2021). Development and trends of intelligent express delivery vehicle technology from JD Logistics practice. *Logistics Technology and Application, 26*(3), 126–128.

Zhao, J. (2020). *Research on multi-floor autonomous navigation technology and system for mobile manipulator robots* (Doctoral dissertation). Harbin Institute of Technology.

Lihuan Wang is a senior physics teacher in Sanfan Middle School and a backbone physics teacher in Beijing. In teaching with their own personality, behavior to infect the students, concerned about the overall growth of students. She advocates that students focus on the practical connections between life and the classroom.

Conclusion: Fostering Youthful AI Ingenuity for a Brighter Tomorrow

The pages of "Youthful Ingenuity" have unveiled a remarkable narrative, one that underscores the transformative role of China in the global AI landscape. Through the lens of the Yuanzhuo Initiatives, this book has brought to light the extraordinary works of our youth, demonstrating their ability to harness artificial intelligence for solving real-world challenges across diverse sectors such as healthcare, education, agriculture, and transportation.

The Yuanzhuo Initiatives, pioneered by Beijing Normal University, have created a unique space where academic institutions, research bodies, and industry converge to nurture young AI talents. The outcome of these initiatives extends far beyond innovative projects; it represents a fresh paradigm in AI education, championing practical, real-world applications.

As we conclude this journey, let us summarize the key takeaways from "Youthful Ingenuity."

1. Youthful Potential Unleashed: The works showcased within these pages are a testament to the creative genius, boundless curiosity, and unwavering determination of our youth. They have harnessed AI to address pressing challenges, proving that age is no bar to innovation. Their Youthful Ingenuity is a wellspring of potential for our society.
2. Practical AI Applications: These innovative projects underscore the fact that AI is not merely theoretical but a practical tool with extensive applications. Ranging from healthcare to transportation, the adaptability of AI is on full display. Stakeholders from diverse sectors can leverage AI to address complex issues.
3. AI Education Revolution: AI education has undergone a transformation, emphasizing real-world experiences. The experiences of these young innovators underline the importance of moving beyond theory to practical applications. This shift can rejuvenate AI education, preparing future generations for a dynamic technological world.
4. Policy and Support: Policymakers and government entities should heed the call presented in this book. Robust policies that support AI education are essential to

R. Huang et al. (eds.), *A Collection of AI Innovations by Chinese Teenagers*, Smart Computing and Intelligence, https://doi.org/10.1007/978-981-97-3316-3

equip the next generation for the AI-driven future. Cultivating environments that encourage innovation among youth is an investment in a brighter tomorrow.
5. Collaboration and Investment: Industry stakeholders should see these cases as potential areas for collaboration and investment. Partnering with these young minds can lead to the development of groundbreaking technologies, fostering economic growth and societal advancement.

Future Directions:

1. Expanding Initiatives: We must strive to broaden the reach of initiatives like the Yuanzhuo Initiatives, ensuring that more youths can participate and innovate. Widespread access to AI education will empower the next generation to fully unlock AI's potential.
2. Interdisciplinary Collaboration: Encouraging greater collaboration among academia, government, and industry can create a nurturing ecosystem for AI education. Cross-disciplinary efforts can remove barriers and accelerate innovation.
3. Diversity and Inclusion: Embracing diversity within the field of AI is essential. Inclusive solutions can cater to the needs of all segments of society. Encouraging a diverse range of voices and perspectives will lead to more comprehensive and equitable AI applications.

In essence, "Youthful Ingenuity" stands as a beacon lighting the way toward a future where artificial intelligence is an integral tool in addressing some of the world's most pressing issues. It is our collective responsibility to ensure that the flame of ingenuity continues to burn brightly and inspires future generations to advance innovation and betterment. The journey of youth-driven AI ingenuity is an invitation for all stakeholders to join hands and create a brighter, more innovative future.

AIED UNESCO CHAIR ON AI IN EDUCATION unesco Chair

UNESCO has approved the establishment of a UNESCO chair on Artificial Intelligence in Education at Beijing Normal University in November 2022. The chair will investigate the use of artificial intelligence in education as a driver and component of solutions and strategies to support the attainment of the sustainable development goals (SDGs).

The goal of the chair is to foster an integrated system of education-related artificial intelligence research, training, information, and documentation. It will facilitate collaboration between high-level, internationally recognized researchers and teaching staff of the university and other institutions in China, as well as elsewhere in Asia, Africa, Latin America, and in other regions of the world.

The chair works on the missions to:

- Improve the capabilities of educators and education managers to use artificial intelligence (AI) ethically and effectively in education.
- Promote international policy dialogue with an emphasis on artificial intelligence in education and the utilization of AI to enable digital transformation in education.

- Facilitate technical innovation on trustworthy, safe, transparent, explainable, and robust AI algorithms, AI tools, and evidence-based, effective models of employing AI in education.
- Support the development of AI talents and high-level graduates of higher education institutions.
- Cooperate closely and actively with concerned entities of UNESCO and existing chairs and networks of UNESCO on relevant programs and activities.

Prof. HUANG Ronghuai who has led research on smart learning environments, artificial intelligence in education, educational technology, and knowledge engineering, takes on the role of Chairholder

北京师范大学智慧学习研究院
Smart Learning Institute of Beijing Normal University

Beijing Normal University (BNU) grew out of the Education Department of Imperial University of Peking established in 1902, which initiated teacher training in China's higher education. After the development for over a century, BNU has become a comprehensive and research-intensive university with its main characteristics of basic disciplines in sciences and humanities, teacher education and educational science. Smart Learning Institute (SLI) is jointly established by Beijing Normal University and a global educational technology company NetDragon Websoft. SLI is a comprehensive experimental platform involving scientific research, technology development, and innovative instruction. SLI focuses on detecting learning patterns powered by ICT, creating smart learning environments and platforms for life-long and life-wide learning, as well as supporting diversified, personalized and differential learning needs for digital learners.

The manufacturer's authorised representative in the EU is Springer Nature Customer Service Centre GmbH, Europaplatz 3, 69115 Heidelberg, Germany. If you have any concerns regarding our products, please contact ProductSafety@springernature.com

Printed and bound by CPI Group (UK) Ltd, Croydon, CR0 4YY

01/10/2026

02225919-0001